Sharp Series

Math 30 Diploma Prep
Pre-Calculus 12

Volume 1
Lessons, Reference
Conceptual Learning

Lex Sharp

Fields of Code Inc.
Calgary, Alberta
www.fieldsofcode.com

Copyright © 2022 Fields of Code Inc.
All Rights Reserved.

Published by Lex Sharp

Art Editor: Bridget Koteles
Cover Image: OpenClipart-Vectors

Fields of Code Inc.
Calgary, Alberta
Canada
www.fieldsofcode.com

No part of this publication may be reproduced in any form or by any means, including scanning, photocopying, or otherwise without prior permission of the copyright holder.

ISBN: 979-8-768-746032

FIRST EDITION

Table of Contents

Preface 1

Lesson 1: Numbering Quadrants 3

Lesson 2: Solving Simple Quadratics via Square Roots.......... 4

    Special Case: Solving for y Instead of x........................... 5

    When to Use ± ? 7

Lesson 3: Solving Quadratic Equations via Factored Forms 10

Lesson 4: The Quadratic Formula 12

    Properties of the Discriminant Δ 13

Lesson 5: Complete the Square Method 14

Lesson 6: Vertex Coordinates, Vertex Form, and Range Definitions.. 19

    The Vertex and Range Definitions 20

Lesson 7: Squaring May or May Not Get Rid of the Square Roots of an Equation.. 21

Lesson 8: The Up/Down Arms of Parabolas 24

Lesson 9: The y-Intercept of a Function 26

Lesson 10: Sketching Graphs, General Approach 27

Lesson 11: Sketching Polynomial Functions 29

Lesson 12: Multiplicity of Roots.. 30

    Drawing Inflection Points 31

Lesson 13: Domain Restriction of Square Roots................ 33

Lesson 14: Sketching Square Roots Below and Above the Line y = x 37

Lesson 15: All Transformations... 39

    Generic Transformations Formula 39

    Mapping Notation 39

    Order of Transformations... 41

Horizontal Variations and Brackets in Function Formulas 42

    Formula vs. Mapping Notation, Correlation 43

Lesson 16: Parameters Substitutions in Base Functions 46

Lesson 17: In Depth Reflections.. 53

Lesson 18: Generating an Inverse... 62

    Example 1 62

    Example 2 65

Lesson 19: Factoring Quadratics Based on Sum and Product of Roots.. 69

Lesson 20: Inequalities with Functions............................... 72

    Typical Inequalities 79

    Techniques Used in Solving Inequalities 80

Algebraic Rule 1 85

Algebraic Rule 2 85

Algebraic Rule 3 85

Algebraic Rule 4 86

Algebraic Rule 5 86

Algebraic Rule 6 86

Algebraic Rule 7 87

Algebraic Rule 8 87

Algebraic Rule 9 87

Algebraic Rule 10 88

Algebraic Rule 11 88

Algebraic Rule 12 88

Other Books in This Series 89

Lessons, Math 30 Diploma Prep

Preface

The series walks learners gradually and conceptually through an understanding of the Math 30 theory through fully explained lessons and solutions to problems. The solutions were organized in smaller manageable booklets representing several unit-tests. These allows students to zero into their wanted topics.

To keep the problem solving volumes concise and affordable, references to theory, summaries and formulas were bundled into individual volumes such as this.

Answers that need to point back to their respective theories use the related volume and lesson number. There the theory can be more thoroughly explained and removes the need to elaborate the same concepts repeatedly over many solutions.

Questions in the Unit Test volumes were chosen for necessary skills needed to solve Math 30 Diploma Exams quickly and easily.

Desmos was used to help demonstrate specific elements of graphs and their transformations.
Find Desmos at: https://www.desmos.com/calculator.

Lex Sharp

Lessons, Math 30 Diploma Prep

Lesson 1: Numbering Quadrants

There are four quadrants trapped between the x and y axes. The first quadrant is positioned over the positive direction of the x-axis and the positive direction of the y-axis. From there numbering happens in a counterclockwise direction.

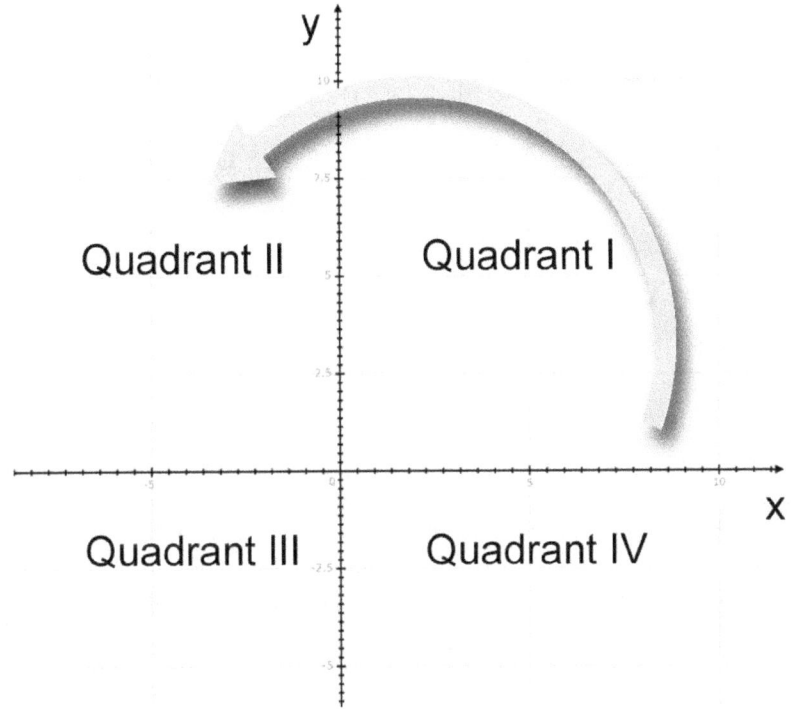

Lesson 2: Solving Simple Quadratics via Square Roots

Imagine the following quadratic equation $x^2 = 1$ and solving for x. The equation is then equivalent to the question: what are the values of x that yield 1 when squared? Visually this looks like so.

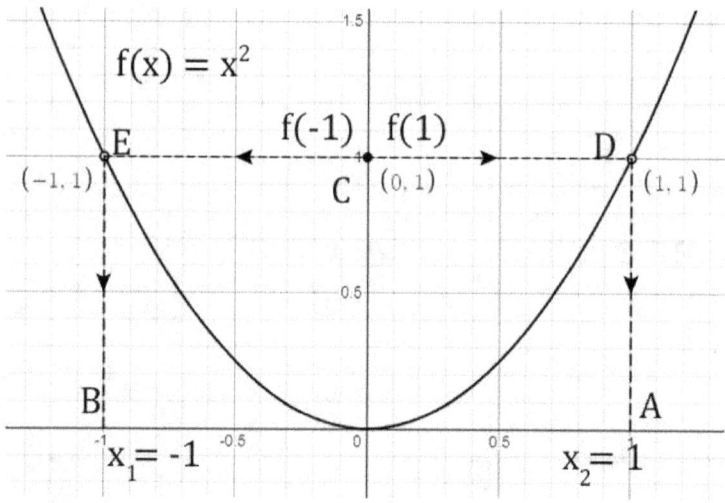

Figure 2.1

It is easy to notice that $x = 1$ solves the equation because $1^2 = 1$. However, 1 is not the only number that works here, -1 works as well.

The function $y = x^2$ represents a parabola. Parabolas are always symmetric. In other words, when looking at a specific y, for example $y = 1$ at point C, the point backtracks to two possible positions on the curve representing the parabola: points D and E. These points are symmetric in relation to the parabola's axis of symmetry, and both D and E reside at the same height $y = 1$. From there, points D and E can further lead to their respective x values:

- point A: $x = 1$ projecting from D,
- point B: $x = -1$ projecting from E.

Both values $x = 1$ and $x = -1$ produce the same height $y = f(x) = 1$

and are both solutions, a.k.a. roots or zeroes of the equation

$$x^2 = 1.$$

This idea can be generalized as follows, if

$$x^2 = number.$$

then there are two possible symmetric solutions for x:

$$x = \pm \sqrt{number}.$$

In an equivalent notation, the two roots are spelled out apart:

$$x_1 = +\sqrt{number}, \text{ and}$$

$$x_2 = -\sqrt{number}.$$

These were the two symmetric points A and B in Figure 2.1, and square rooting worked because the square-root is the opposite operation of the squared exponent.

One important thing to remember is that solving for x is identical to isolating x.

Additionally, isolating x^2 is not the same as isolating x. In other words, x^2 cannot be the solution, rather x is. Initial attempts to solve the quadratic will typically end up with isolating x^2 and once this is accomplished, the goal becomes to remove the square exponent from x via a square root operation so that x can stand alone isolated in the $x = \pm \ldots$ form.

This simple case can be extended to more complicated examples described in the special cases below.

Special Case: Solving for y Instead of x.

Rather than solving for x, solving for y can be an alternate goal sometimes. For example, it is assumed that two expressions are given:

 i. *expression1* is based on y, and

ii. *expression2* is based on x,

and together they form a quadratic equation as shown here,

$$(expression1)^2 = expression2$$

but this time the equation must be solved for y.

An example clarifies this:

$(y + 8)^2 = x^2 - 4x + 4$, and the requirement is to solve for y.

Solving for y means isolating y. However, the square exponent on the left side interferes with the isolation of y. Instead, initially only $(y + 8)^2$ can be isolated, and there is no access to the number 8 until the square is removed. To rid the left side of the square exponent, square rooting is applied on both sides of the equation. This is done accounting for the symmetry via the $\pm$ signs that was explained earlier. Then,

$$(expression1)^2 = expression2 \text{ becomes}$$

$$expression1 = \pm \sqrt{expression2},$$

$$y + 8 = \pm \sqrt{x^2 - 4x + 4}.$$

This opens access to the number 8 which brings the equation much closer to the isolating of y. The next flow helps track the solution till the end.

$y + 8 = \pm \sqrt{x^2 - 4x + 4}$

$y + 8 - 8 = \pm \sqrt{x^2 - 4x + 4} - 8$ 8 is removed from both sides, with the goal to isolate y.

$y = \pm \sqrt{x^2 - 4x + 4} - 8$

$y = \pm \sqrt{(x - 2)^2} - 8$ The expression in x under the square root (right side) happens to be a perfect square (*see Algebraic Rule 11*). If this wasn't

$y = \pm (x - 2) - 8$ the case, this would be the last step and the solution.

The square root and the square exponent annihilate each other. The $\pm$ sign is not affected.

$y_1 = +(x - 2) - 8 = x - 10$

$y_2 = -(x - 2) - 8 = -x - 6$

Two possible roots y_1 and y_2 were derived from the $\pm$ variations.

The two roots for y that solve $(y + 8)^2 = x^2 - 4x + 4$ are the following expressions in x:

$$y_1 = x - 10, \text{ and}$$

$$y_2 = -x - 6.$$

Did You Know?

The scenario described above is commonly seen in the calculations of **Inverse** formula definitions. Read more about the Inverse in *Lesson 18*.

When to Use $\pm$?

When starting out with an equation stated as a parabola (in the previous example that was x^2), there are two possible solutions based on $\pm$ due to symmetry of the parabola. See an additional example next.

$x^2 = 9$

$x = \pm \sqrt{9}$

$x = \pm\ 3.$

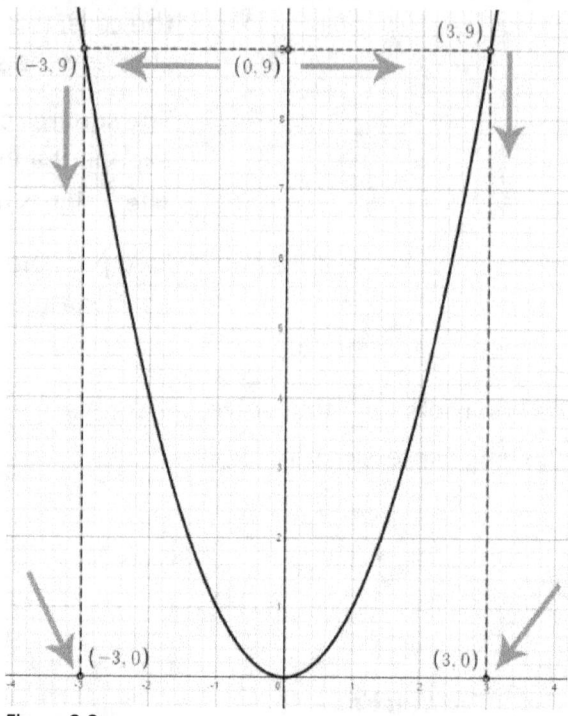

Figure 2.2

The starting point of the equation is a parabola. This reenforces the need for using two symmetric values as roots, one on either side of the axis of symmetry. Notice $y = 9$ backtracks to two possible x values: 3 and -3. This is the same as explained *in Figure 2.1* earlier.

In contrast, when starting out with a square root function, there is only one positive solution. The square root function is not symmetric and dual signs $\pm$ are **not** applicable. For example this is the case for:

$$\sqrt{9} = 3,\text{ or generally:}$$

$$\sqrt{x} = a,\text{ where } a \text{ is positive.}$$

There is only one solution because $y = 3$ backtracks to only one possible value: 3, as shown in *Figure 2.3*. Only one x value produces the wanted $y = 3$.

Lessons, Math 30 Diploma Prep

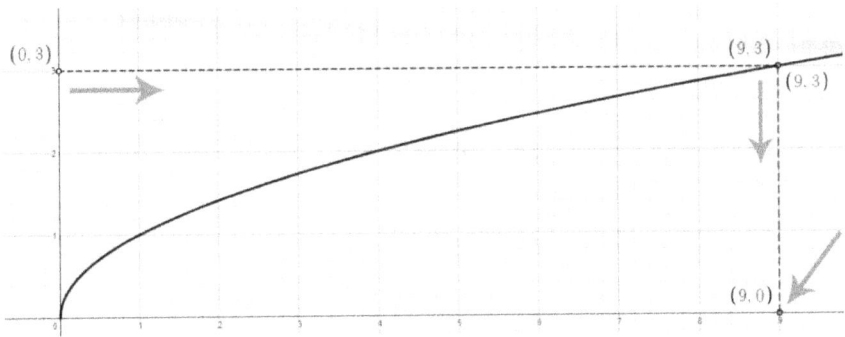

Figure 2.3

In summary, even though the ± and the square root are often seen together, the ± can only originate from an equation based on a parabola, never from one based on a square root function.

In Conclusion...

$x^2 = number$	$\sqrt{x} = a$, where a is positive
Origin is a parabola.	Origin is a square root function.
Has **two** answers with ± for x.	Has only **one** positive answer for x.

Lesson 3: Solving Quadratic Equations via Factored Forms

Imagine an equation $f(x) = 0$ and the expression $f(x)$ can be factored into two terms.

Imagine that each factoring term is represented by its own function, one called $g(x)$ and the other $h(x)$. Then $f(x) = 0$ is equivalent to

$$g(x) * h(x) = 0.$$

The clever part that makes factoring useful is that there is only one way these two can produce a zero when combined into a product: one or the other term must be zero. That is either

(Option 1) $g(x) * 0 = 0$ makes the zero happen, or

(Option 2) $0 * h(x) = 0$ does.

The above two ideas are the basis for the equation's two possible solutions, i.e.:

$h(x) = 0$, in Option 1, or

$g(x) = 0$, in Option 2.

It can be said that the two expressions take turn at being zero. Each contributes an individual solution to the equation. In short: either $g(x)$ or $h(x)$ must be zero for the product to be zero.

This is the motivation for factoring functions.

Additionally the terms $g(x)$ and $h(x)$ will result in lower exponents which makes them easier to solve and replace the need to deal with the more complicated $f(x)$.

Therefore, when solving the equation $f(x) = 0$ with a factored form of $f(x) = g(x) * h(x)$

(i) the first solution x_1 is donated by $g(x) = 0$,
(ii) the second solution x_2 is donated by $h(x) = 0$.

This is best clarified through an example:

$$x^2 - 3x + 2 = 0,$$

is equivalent to the factored form:

$$(x - 1)(x - 2) = 0.$$

The first solution is donated by

$$g(x) = x - 1 = 0, \text{ therefore } x_1 = 1.$$

The second solution is donated by

$$h(x) = x - 2 = 0, \text{ in this case } x_2 = 2.$$

Then the two roots that solve the $x^2 - 3x + 2 = 0$ equation via the factoring

$$(x - 1)(x - 2) = 0,$$

are $x_1 = 1$ and $x_2 = 2$.

Lesson 4: The Quadratic Formula

Quadratic equations can be solved through several methods. For example, one common way to solve is factoring, and is considered the mathematically elegant solution (see *Lesson 3*).

However, factoring is not always possible.

The quadratic formula can be used instead. The quadratic formula is not considered the most elegant solution but is accurate and provides a failsafe when all other methods fail. The method assumes the quadratic equation is already arranged in **standard form**, with a single term for each exponent, sorted in descending order of the exponent (descending from left to right), like so

$$ax^2 + bx + c = 0.$$

Then based on the sorted terms' coefficients, the following ideas can help solve for x:

$$\Delta = b^2 - 4ac,$$

Δ is known as the **Discriminant**.

The roots of this equation are calculated as

$$x_{1,2} = \frac{-b \pm \sqrt{\Delta}}{2a}.$$

This formula represents two possible roots and can alternately be stated as

$$x_1 = \frac{-b + \sqrt{\Delta}}{2a},$$

and

$$x_2 = \frac{-b - \sqrt{\Delta}}{2a}.$$

Properties of the Discriminant Δ

The discriminant exists as its own concept in Math because it helps "discriminate" between quadratic equations that have either 0, 1 or 2 distinct roots.

There is a correlation between the sign of Δ and the number of roots following these rules:

$\Delta < 0$ If the discriminant is negative, there are no real roots for the given quadratic equation. Thus, x_1 and x_2 do not exist.

When the discriminant is negative, the formula for the roots $x_{1,2}$ cannot enforce the square root for the $\sqrt{\Delta}$ element, because square root function can only operate on positive numbers, it cannot produce an answer for a negative number.

$\Delta = 0$ If the discriminant is zero, then the given quadratic equation has two identical roots: $x_1 = x_2$, often referred to as rather just one root.

$\Delta > 0$ The given quadratic equation has two distinct roots: $x_1 \neq x_2$.

Lesson 5: Complete the Square Method

The **Complete the Square** method rearranges a quadratic of the form

$$x^2 + bx + c$$

into a "perfect square plus an imperfection" so that the imperfection in its arrangement is always a constant.

Additionally, the "complete the square" model is equivalent to the **vertex form** of the given quadratic. The arrangement is shown below.

$$x^2 + bx + c = \underbrace{\left(x + \frac{b}{2}\right)^2}_{\text{perfect square}} + \underbrace{c - \left(\frac{b}{2}\right)^2}_{\text{and imperfection}}$$

Rather than memorizing the above formula, a method is available for converting a quadratic into a "complete the square" expression based on the following steps.

Step 1

When working with the quadratic $x^2 + bx + c$, b is expressed as twice the half of b, i.e.

$$b = 2\left(\frac{b}{2}\right).$$

This new expression still equals b. It is simply rewritten as an equivalent form. This new expression is plugged back instead of the original b like so,

$$x^2 + bx + c = x^2 + 2\left(\frac{b}{2}\right)x + c.$$

Step 2

Just like in Step 1, various other equivalent expressions can be used replace some of the original terms. The following are such quirks and are purposefully introduced into the quadratic expression instead of various existing terms:

(i) adding a zero to an expression does not change its value, i.e. $z = z + 0$,
(ii) expressing zero as any number minus itself can be used instead of a zero based on $0 = w - w$.

Zero is expressed as the square of the previously introduced half b, minus this same expression as shown:

$$0 = \left(\frac{b}{2}\right)^2 - \left(\frac{b}{2}\right)^2,$$

this *zero* is then added to the expression accomplished so far:

$$x^2 + bx + c = x^2 + 2\left(\frac{b}{2}\right)x + c + 0$$

$$= x^2 + 2\left(\frac{b}{2}\right)x + c + \left(\frac{b}{2}\right)^2 - \left(\frac{b}{2}\right)^2.$$

Step 3

There are a few terms in the expression in Step 2 that can already be bundled to form a perfect square based on *Algebraic Rule 11*:

$$(a + b)^2 = a^2 + 2ab + b^2.$$

The expression terms that form the perfect square are emphasized below:

$$x^2 + bx + c = x^2 + 2\left(\frac{b}{2}\right)x + c + \left(\frac{b}{2}\right)^2 - \left(\frac{b}{2}\right)^2.$$

The emphasized perfect square terms are separated from the rest in brackets for more clarity:

$$x^2 + bx + c = \left(x^2 + 2\left(\frac{b}{2}\right)x + \left(\frac{b}{2}\right)^2\right) + c - \left(\frac{b}{2}\right)^2.$$

Step 4

The separated terms that were placed in the brackets, are replaced by the perfect square. The remaining terms are bundled to form a constant representing **an imperfection** to the said perfect square.

$$x^2 + 2\left(\frac{b}{2}\right)x + \left(\frac{b}{2}\right)^2$$

$$x^2 + bx + c = \left(x + \frac{b}{2}\right)^2 + c - \left(\frac{b}{2}\right)^2.$$

In conclusion, the "complete the square" formula of any quadratic $x^2 + bx + c$ is an arrangement made of

(i)   a perfect square binomial, plus
(ii)   an imperfection made of only constants,

as shown below.

$$x^2 + bx + c = \underbrace{\left(x + \frac{b}{2}\right)^2}_{\text{perfect square}} \text{ and } \underbrace{c - \left(\frac{b}{2}\right)^2}_{\text{imperfection}}$$

In other words, even a quadratic that is not a perfect square, is only one constant away from a perfect square.

Example, Complete the Square

$$x^2 + 6x + 2 =$$

$$x^2 + 2\left(\frac{6}{2}\right)x + 2 =$$ Step 1 is applied.

$$0 = \left(\frac{6}{2}\right)^2 - \left(\frac{6}{2}\right)^2$$ Step 2 is applied. A special zero is created and added to the expression, this is without changing its original value.

$$x^2 + 2\left(\frac{6}{2}\right)x + 2 + 0 =$$

$$x^2 + 2\left(\frac{6}{2}\right)x + 2 + \left(\frac{6}{2}\right)^2 - \left(\frac{6}{2}\right)^2 =$$

$$x^2 + 2\left(\frac{6}{2}\right)x + 2 + \left(\frac{6}{2}\right)^2 - \left(\frac{6}{2}\right)^2 =$$ Step 3 is applied. Terms of an identifiable perfect square are highlighted based on *Algebraic Rule 11*.

$$\left(x + \frac{6}{2}\right)^2 + 2 - \left(\frac{6}{2}\right)^2$$ Step 4 is applied to show a perfect square as part of the original expression and is followed by constants only which form an imperfection to the square.

The expression can be further beautified as necessary. The only important thing now is to not reopen the brackets, to leave the square on, and to maintain the "perfect square **plus an imperfection**" form. Multiple constants can be bundle together for a tidy look:

$$\left(x + \frac{6}{2}\right)^2 + 2 - \left(\frac{6}{2}\right)^2 =$$
$$(x + 3)^2 + 2 - (3)^2 =$$
$$(x + 3)^2 + 2 - 9 =$$
$$(x + 3)^2 - 7.$$

The "complete the square form", is thus:

$$x^2 + 6x + 2 = (x + 3)^2 - 7.$$

This can be easily converted into the quadratic's equivalent **vertex form** with one more tweak (read more about **Vertex Form** in *Lesson 6*). The tweak is to format the above expression to show a subtraction within the squared binomial rather than an addition.

In general, the vertex form shows the subtraction like so

$$f(x) = a(x - p)^2 + q,$$

i.e. the p component must appear accompanied by a minus sign. This is easy to do in the current example since the $+3$ in the binomial $(x + 3)$ can be written as:

$$+ 3 = - (-3),$$

then

$$x + 3 = x - (-3),$$

$$(x + 3) = (x - (-3)).$$

Thus, after plugging in the new expression equivalent to $+ 3$, we get the final vertex form:

$$x^2 + 6x + 2 = (x - (-3))^2 - 7, \text{ or}$$

$$x^2 + 6x + 2 = (x - (-3))^2 + (- 7),$$

which overlaps with the formula

$$f(x) = a(x - p)^2 + q.$$

This helps identify the vertex (see *Lesson 6*) to be the point at these coordinates

$$V(p, q) = V(-3, -7).$$

Most often when quadratic problems require a vertex form, it is calculated via the **complete the square** method.

The very purpose of completing the square for a quadratic is to find the vertex form or for use in the calculation of an inverse which will be discussed further in *Lesson 18*.

Lesson 6: Vertex Coordinates, Vertex Form, and Range Definitions

The wording "parabola" and "quadratic" are interchangeable. These phrases will be used below.

The **vertex** is the minimum or the maximum point of a quadratic function, and all quadratics have one such.

The following is a generic parabola and is shown in standard form:

$$f(x) = ax^2 + bx + c. \quad \textit{(Expression 6.1)}$$

Its vertex can be expressed as $V = (x_v, y_v)$ using the formula:

$$x_v = -\frac{b}{2a}.$$

The y coordinate is also available through a formula:

$$y_v = c - \frac{b^2}{4a}$$

but does not need to be memorized since it can be derived by plugging the value for x_v into $y = f(x)$. This way the y value can be calculated from there:

$$y_v = f(x_v) = f\left(-\frac{b}{2a}\right).$$

All coefficients a, b and c belong to the above quadratic standard form (see *Expression 6.1*). The formulas shown above for x_v and y_v are rarely used. Instead most solutions rely on the **Vertex Form** of quadratics, which is described next.

All quadratic formulas can be molded into the **vertex form** via the "complete the square" method as described in *Lesson 5*. The vertex form is depicted as

$$f(x) = a(x - p)^2 + q.$$

Next, the coefficients p and q can be picked out of this form to express the actual vertex coordinates of the parabola at: $V(p, q)$.

The advantage of working with the vertex form is that it produces an arrangement that allows a very quick identification of the vertex coordinates $V(p, q)$.

The Vertex and Range Definitions

The vertex is useful when sketching functions and when determining a function's range. This is so because the vertical line that travels through a vertex is the parabola's axis of symmetry. The vertex is either a maximum or minimum.

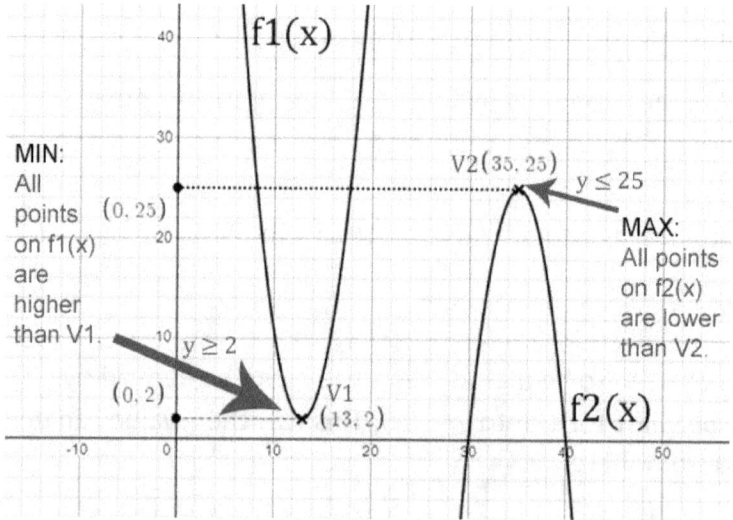

This is how this fact is used, the vertex is the

(i)   lowest y point when the vertex is a minimum (for example $V1$ in the figure above), or

(ii)   the highest y point when the vertex is a maximum (for example $V2$ in the figure above).

The y values of the **Range** are formally defined as:

(i)   the range is all the y values that are smaller than the maximum point (see $V2$ above): $y \leq p$ a.k.a. $y \in (-\infty, q]$, or

(ii)   the range is all y values that are larger than the minimum point (see $V1$ above): $y \geq p$ a.k.a. $y \in [q, +\infty)$.

Lesson 7: Squaring May or May Not Get Rid of the Square Roots of an Equation

Whenever x appears under a square root in an equation, we prefer to get rid of the square root so that x can be isolated and thus solved for. The next examples refer to such equations and contrast the two possible cases:

(i) a square root **is discarded** via squaring, and
(ii) squaring **fails to discard** a square root.

The squaring in question is applied to both sides of the equation.

Case (i):

The expression

$$\sqrt{x+1} = \cdots$$

resides alone on one side of an equation, and the other side had no occurrences of x. Thus x is partially isolated on the left.

Both sides of the equation are about to be squared. The result of the squaring on the said side yields:

$$(\sqrt{x+1})^2 = (\ldots)^2.$$

The square root and the square annihilate each other and result in:

$$x + 1 = (\ldots)^2.$$

Thus, this operation gets rid of the square root that was associated with the x.

Case (ii):

One side of an equation contains a squared root expression of x and is also accompanied by a constant outside of the squared root, in the next example that constant is the number 4:

$$\sqrt{x+1} + 4 = \ldots$$

Typically the hope is to get rid of the square root so that x can be isolated, which then leads to x being solved for.

A squaring of both sides of this equation can proceed in one of two options:

- opening the brackets of $(\sqrt{x+1} + 4)^2$ and bundling like-terms together after applying the *FOIL* method,
- applying *Algebraic Rule 11*: $(a+b)^2 = a^2 + 2ab + b^2$.

The method shown next is the latter. The squared root element is emphasized so that it can be tracked. Check if the square root has indeed been removed.

$$(\sqrt{x+1} + 4)^2 =$$
$$= \left(\sqrt{x+1}\right)^2 + 2 * \sqrt{x+1} * 4 + 4^2 =$$
$$= (x+1) + 8 * \sqrt{x+1} + 16.$$

This operation did not get rid of the square root. $\sqrt{x+1}$ remains present in the middle term.

Using the constant 4 in the previous example was the simplest case. Other examples involving even more complicated terms instead of the constant 4 can happen and result in the same limitations, for example:

$$\sqrt{x+1} + 4x = ...$$

Conclusion:

Whenever a square root is **by itself** on one side of an equality, such as it is in *Case (i)*, and it is squared, the squaring has an eliminating effect on the square root. For example,

$$\left(\sqrt{a}\right)^2 = a.$$

However, if the square root is **not by itself** but rather followed by an additional constant or some other term as it is in *Case (ii)*, then squaring will fail to remove the square root.

This is because during the evolving of the terms of the squaring using *FOIL* or *Algebraic Rule 11*, the middle term continues to carry a square root factor as demonstrated below:

$$(\sqrt{a} + b)^2 = (\sqrt{a})^2 + 2 * \sqrt{a} * b + b^2$$

$\sqrt{a}$ is the square root intended for removal via the squaring.	The $(\sqrt{a})^2$ element in the result gets rid of the square root via squaring. This particular term works well as planned. However this is not enough.	The middle term $2*\sqrt{a}*b$ is the one causing the trouble. It continues to hold on to the square root $\sqrt{a}$ factor. The squaring did not get rid of the offending square root, as shown in *Case (ii)* earlier.

In conclusion, when attempting to get rid of a square root in an equation, the square root element needs to be completely isolated on one side of the equation by itself, with no additional constants or other terms. Otherwise the squaring will be futile and will not rid the expression of its square root.

By this conclusion the example in *Case (ii)* can be solved by subtracting 4 from both sides of the equation as shown below, which transforms it into a *Case (i)* pattern which is easy to solve.

$$\sqrt{x+1} + 4 = \ldots$$

$$\sqrt{x+1} + 4 - 4 = \ldots - 4$$

$$\sqrt{x+1} = \ldots - 4$$

Lesson 8: The Up/Down Arms of Parabolas

The leading coefficient of a parabola correlates with the directions of the parabola's arms. If the parabola is expressed in standard form: $f(x) = ax^2 + bx + c$, then the leading coefficient is a.

If the parabola is not given in standard form, the terms must be rearranged by opening brackets, bundling like-terms together and so on, until the standard form is obtained.

When the leading coefficient a is

   I. **positive**, then the arms go **up**.
   II. **negative**, then the arms go **down**,

and examples are shown in the plots below.

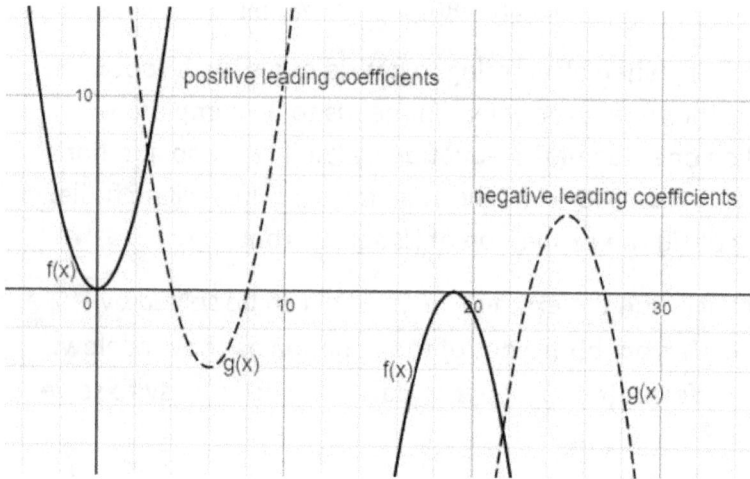

Figure 8.1

The next graphs turn the words **positive** and **negative** as follows to help remember the direction of the arms depending on the values of a. This is a well-known technique used by many teachers.

$a > 0 \rightarrow$ **positive** $\rightarrow$ think of a **smiley** attitude, and
$a < 0 \rightarrow$ **negative** $\rightarrow$ think of a **frowny** attitude.

The sign of a, the leading coefficient, is synched with the "attitude" of the parabola to remember. In this model the smile/frown is in sync with the curvature of the parabola. That is smiley is with the arms up, and frowny is with the arms down as shown in the figure below.

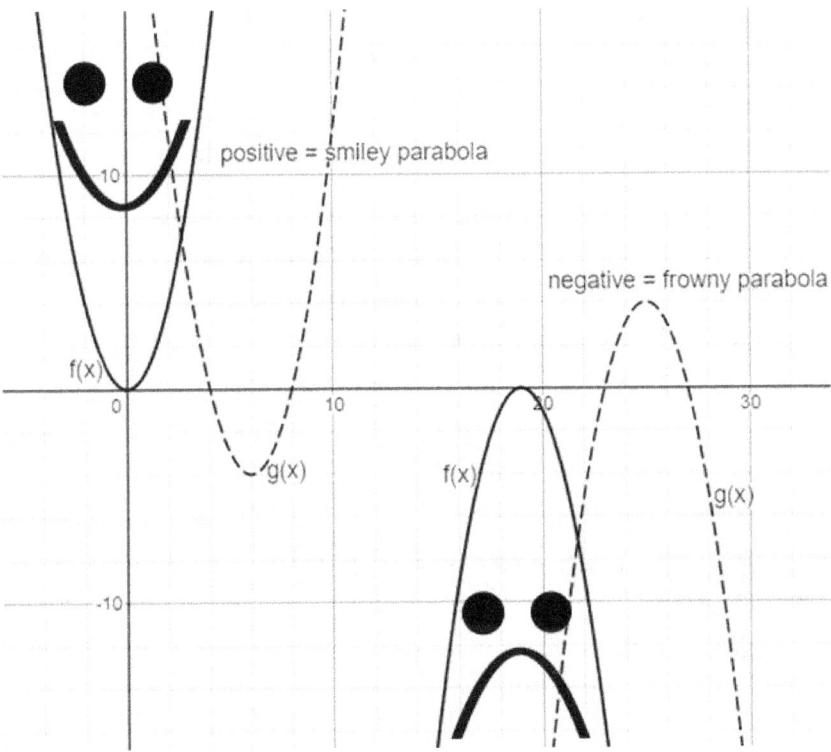

Lesson 9: The y-Intercept of a Function

The y-intercept of any function $f(x)$ is determined by plugging in the value of $x = 0$ to get the intercept $y_i = f(0)$.

Imagine a function to be of any power shown in standard form:

$$f(x) = a_1x^n + a_2x^{n-1} + \ldots + a_{n-1}x^2 + a_nx + a_{n+1}$$

Substituting *zero* will neutralize all terms that have x in them because each such term turns into *zero* when multiplied by $x = 0$.

The only unaffected terms are those that do not have x in them. In the above arrangement there is only one such term: the trailing constant a_{n+1}. It's the only one to not have x which is why it is not zeroed away. In the case of the quadratic function $f(x) = ax^2 + bx + c$, the only such term is c, thus c equals the y-intercept.

In conclusion, for any function $f(x)$:

$f(0) = a_{n+1}$ = *the trailing constant of the function* = *the y-intercept*.

The y-intercepts are Cartesian points of the form $(0, y_i)$, where y_i is the y coordinate at which the parabola intercepts the y-axis and is equivalent to the constant term of the function $y_i = a_{n+1}$.

The beauty of this method is that if the function is already arranged in standard form, there is no need for calculations. The constant that equals the y-intercept is simply picked out from the function formula. Furthermore, if the function is not so arranged, all that is needed is to bundle all constants of the function formula that are free of x into a single term without having to rearrange the rest, which is a minimal effort.

$$f(x) = a_1x^n + a_2x^{n-1} + \cdots + a_{n-1}x^2 + a_nx + \boxed{a_{n+1}}$$

the y-intercept

The y-intercept is necessary for the sketching of functions.

Lesson 10: Sketching Graphs, General Approach

Plotting a function manually, a.k.a. **sketching**, is done based on several elements that need to be tracked down or calculated before the plotting proceeds.

The following elements are typically examined while sketching a graph:

1. The positive sides of the two axes are marked as x for the horizontal axis and as y for vertical axis.
2. The type of graph is established based on the highest powered term which is called the **leading term** of the function formula. The type can be line, quadratic, cubed, or a plot of higher exponents.
3. The range is considered during plotting and often the expression of the range is jotted beside the graph.
4. Any domain restrictions and NPV-s are determined and stated.
5. Singularities or asymptotes are assumed for all NPV-s. Convergence is explored at the extreme areas, known as **end behaviors**. Horizontal asymptotes may also be present.
6. Leading coefficients for leading exponents > 1 are examined to determine how the **outer arms** approach their end behavior (see for example *Lesson 8*):
    (iii) both upwards,
    (iv) both downwards,
    (v) opposite directions, for example it may start in the third quadrant, end in the first quadrant, or start in the second quadrant and end in the fourth quadrant.
7. The x-intercepts, a.k.a. the roots (or zeroes) are calculated based on $f(x) = 0$ (i.e. solving for x) and are marked on the graph.
8. The y-intercept is calculated and marked on the graph (see *Lesson 9*).
9. If the graph is symmetric, the axis of symmetry is drawn to anchor the graph visually. For example, all parabolas have a

vertical line through the x of their vertex which is the function's axis of symmetry. There are other types of functions/formulas that don't behave so. For example, the inverse formula of a quadratic function has a horizontal line as its axis of symmetry rather than vertical. The inverse also has the diagonal $y = x$ as its axis of symmetry in relation to its base function. Not all functions are symmetric.

10. All parabolas have a vertex which represents a minimum or maximum of the function. The vertex should always be included in the sketch of parabolas. Completing the square on the quadratic may be necessary to discover the vertex coordinates (read more about completing the square in *Lesson 5* and the vertex in *Lesson 6*). As a minimum or maximum, the vertex is also significant because it is a point beyond which none of the function's y values go past.

11. If the graph plotted is an inverse, then the line $y = x$ is drawn and acts as an axis of symmetry between the inverse and its base function from which it derives.

12. If transformations are plotted side by side with their base functions, then if invariant points exist, then these are also noted on the graph.

13. If transformations are plotted side by side at least the function name, for example, $f(x)$, $g(x)$, $h(x)$, etc., and often the function formula as well are displayed next to each plot.

Lesson 11: Sketching Polynomial Functions

The leading term of a polynomial carries its leading exponent. The type of exponent: even or odd, has a relationship to visual aspects of the resulting graph.

The following rules apply when sketching a polynomial $p(x)$.

- A leading **exponent of** 1 means that $p(x)$ graphs a straight line.
- If the leading **exponent is** 2, the shape of $p(x)$ is a parabola. A parabola can be seen as an approximation of a rounded letter V. That is the arms of the parabola proceed away from each other infinitely.
- If the **leading exponent is even** and larger than 2, than the graph has its outer arms pointing in the same direction, i.e. either:
    - both upwards if the leading coefficient is positive, or
    - both downwards if the leading coefficient is negative.
- If the **leading exponent is odd** but larger than 1, then the two outer arms of the graph point in opposite downwards and upwards directions.
    - If the leading coefficient is **positive**, then the graph starts in the lower left quadrant at $(-\infty, -\infty)$ and ends up in the top right quadrant at $(+\infty, +\infty)$.
    - If the leading coefficient is **negative**, then the graph starts in the top left quadrant at $(-\infty, +\infty)$ and ends up in the bottom right quadrant at $(+\infty, -\infty)$.

To fully detail a graph, one must first lock in the outer arms of the function as described above. This is followed by marking each root and providing it with the shape related to the root's multiplicity as described in *Lesson 12*. From here the function is "forced" to travel between the outer arms and sequentially through the roots from the left to the right of the horizontal number line, i.e. in ascending order of the roots' appearance on the axes.

Lesson 12: Multiplicity of Roots

The multiplicity of a root is an exponent. It represents how many times a binomial appears in the factored form of a function. For example,

$$f(x) = (x+1)(x-1)^2(x+5)(x-4)^3(x+1)^3,$$

has several roots and each has its own multiplicity.

For this function, the binomials (and thus their roots) in the representing equation

$$(x+1)(x-1)^2(x+5)(x-4)^3(x+1)^3 = 0,$$

have the following multiplicities sorted in increasing order:

Original term	binomial	root	multiplicity = power
$(x+5)$	$(x+5)$	$x = -5$	1
$(x-1)^2$	$(x-1)$	$x = 1$	2
$(x-4)^3$	$(x-4)$	$x = 4$	3
$(x+1)^1(x+1)^3$	$(x+1)$	$x = -1$	4

The multiplicity of a root has only **one of the three** following visual qualities:

- a straight line through the x-axis, or
- a dome tangent to the x-axis (below or above), or
- an inflection point through the x-axis.

The function shown in *Figure 12.1* presents example shapes for the above mentioned multiplicities as follows:

root	shape	multiplicity (**m** in *Figure 12.1*)
$x = 3$	line	1
$x = 1$	dome	even ≥ 2
$x = 5$	inflection	odd ≥ 3

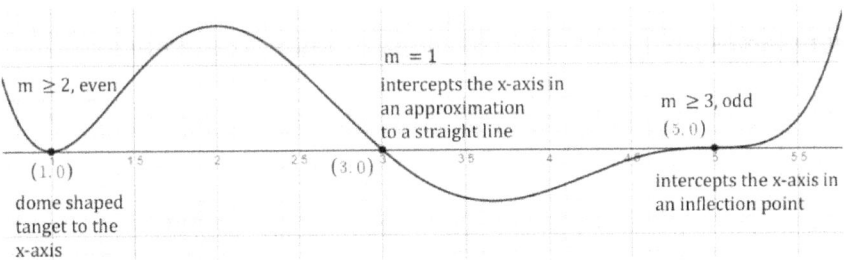

Figure 12.1

Alternately, the mirror image has equivalent shapes from the other side of the x-axis.

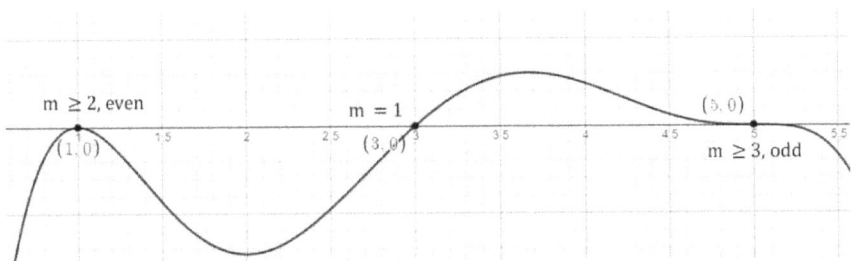

Figure 12.2

These categories are an important visual aspect of all plotted curves. They help sketch correctly around the roots using the following rules:

- Even multiplicities cause dome-like shapes that are tangent to the x-axis. A higher even multiplicity provides a dome with a wider "tangent footprint" with the x-axis.
- A multiplicity of 1 channels the plot line in a straight line through the x-axis cutting straight through its root.
- Odd multiplicities that are larger than 1 generate inflection points.

Drawing Inflection Points

The following graph shows an inflection at point B at coordinates $(5, 0)$.

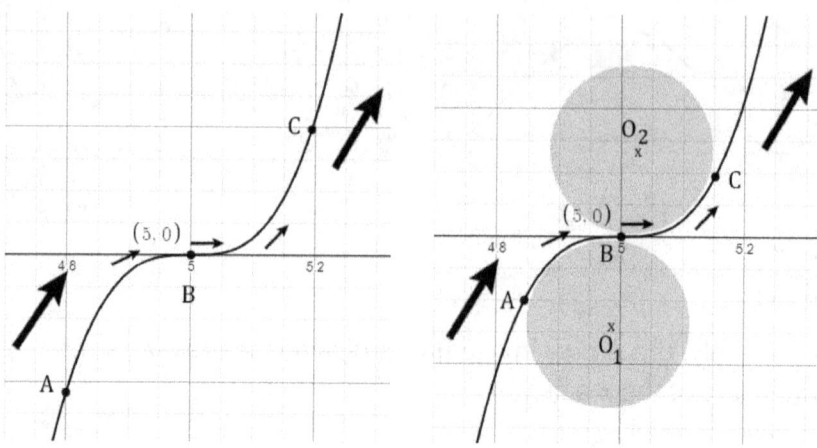

Figure 12.3

There are many ways to characterize the inflection point curve, for example: the line maintains an upwards trend but swerves, or slaloms through $(5, 0)$ before it returns to its original trend.

Another way to look at this shape is to imagine a circle that best approximates the curve near points A and B, with the center at O_1. Considering B is the inflection point, the part of the curve that travels between B and C approximates to a completely different circle centered at O_2, this center is symmetrically opposed to O_1 on the other side of the imaginary line AC, and the other side of the x-axis. Both circles at O_1 and O_2 are:

- tangent to each other at B,
- tangent to the x-axis at B.

For Math 30, all that matters is **to approximate** the curvature roughly as shown from A to B to C, drawing just enough of the "bumps" to clearly indicate on a graph that the multiplicity is odd and ≥ 3, and to ensure such intercept does not travel through a straight line or a dome-like shape.

Lesson 13: Domain Restriction of Square Roots

The square root function $f(x) = \sqrt{x}$ has a domain restriction of

$$\{ x \mid x \geq 0, x \in R \}.$$

The square root function exists in the first quadrant only as shown below. This means no negative x values can be fed to $f(x)$ and no negative y values are ever produced by this function.

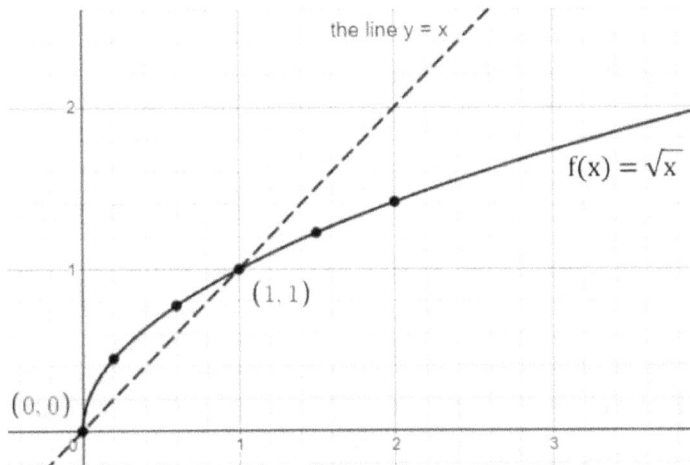

Figure 13.1

If a transformation $g(x)$ is used to horizontally shift $f(x)$ left or right by h units, then the domain of the transformation is also shifted accordingly.

There are two possibilities: a positive or a negative h for a shift scenario, as shown by *Case I* and *Case II* next. These two cases shift by 9 units left and right in the example that follows. The formula used was identical to the one described in *Lesson 15*:

$$g(x) = a * f(\, b(x - h) \,) + k,$$

but since the only changes considered were horizontal shifts, the vertical coefficients k and a are removed, and the formula translates into a much simpler form:

$$g(x) = f(x - h).$$

Notice the arrangement and the importance of having a minus between x and h. In other words h must be subtracted.

Case I $9 > 0$ $g(x) = f(x - 9)$ the shift happens to the **Right** by 9 units,

Case II $-9 < 0$ $g(x) =$ the shift happens to the **Left** by 9 units.
$f(x - (-9)) =$
f(x + 9)

The result as specifically applied to the square root $f(x) = \sqrt{x}$ is:

Case I $g(x) = f(x - 9) = \sqrt{x - 9}$ the shift happens to the **Right** by 9 units,

Case II $g(x) = f(x + 9) = \sqrt{x + 9}$ the shift happens to the **Left** by 9 units.

The **Domain Restriction** in each case can be calculated by stating that whatever resides under the square root, no matter how complicated an expression, needs to be overall positive or zero:

$$\text{expression under square root} \geq 0.$$

The following explanation clarifies this:

the domain of $\sqrt{x - 9}$	is valid only if $x - 9 \geq 0$, $x \geq 9$	This is an inequality in x and thus it must be solved for x regardless of what the expression under the square root is.
The domain of $\sqrt{x + 9}$	is valid only if $x + 9 \geq 0$, $x \geq -9$.	This is an inequality that must be solved for x.

The following two examples drill thoroughly into this concept.

Example 1:

$\sqrt{x-9}$	the domain restriction is dependent on whatever is found underneath the square root, here that is $x - 9$.
the domain of $\sqrt{x-9}$	is valid only if $x - 9 \geq 0$. This is an inequality that must be further solved for x.
	$x - 9 + 9 \geq 0 + 9$ $x + 0 \geq 9$ $x \geq 9$	9 is added to both sides of the inequality to isolate x.

Therefore the solution to the required inequality is $x \geq 9$ and it represents the permissible domain for the function $\sqrt{x-9}$.

Example 2

$\sqrt{x^2 - 9}$	The domain restriction is dependent on whatever is found underneath the square root, here that is $x^2 - 9 \geq 0$.
the domain of $\sqrt{x^2 - 9}$	is valid only if $x^2 - 9 \geq 0$. This is an inequality that must be solved for x.
	It is a very **common mistake** to expect x to be positive rather than the full expression $x^2 - 9 \geq 0$. This inequality is solved below.

Solving the $x^2 - 9 \geq 0$ inequality for x begins with factoring the quadratic.

$x^2 - 9$ is easily factored into a difference of squares (see *Algebraic Rule 8*) because 9 is a perfect square and can be expressed as 3^2.

Therefore $y = x^2 - 9 = x^2 - 3^2 = (x + 3)(x - 3)$.

The inequality $y = x^2 - 9 \geq 0$ is stated in factored form as

$$y = (x + 3)(x - 3) \geq 0.$$

The leading term is $x^2 = 1 * x^2$, thus the leading coefficient is 1 and positive. Therefore, $y = x^2 - 9 = (x + 3)(x - 3)$ is an arms-up parabola (see *Lesson 8*). This function has two distinct roots $x = 3$ and $x = -3$ which are derived from the two factored terms taking terms at being zero (see *Lesson 3*).

This means the inequality $y \geq 0$ has positive y values **"outside the roots"**. To better understand this statement read *Lesson 20*, where the "outside the roots" concept is visualized graphically.

Thus the solution to the inequality $y = (x + 3)(x - 3) \geq 0$ is the range of x outside the roots, i.e.:

$$x \leq -3 \text{ and } 3 \geq x,$$

which is also equivalent to

$$x \in (-\infty, -3] \text{ and } x \in [3, +\infty).$$

This solution in x represents the domain of the given square root function $f(x) = \sqrt{x^2 - 9}$:

$$\text{Domain} = D = \{ x \mid x \leq -3 \text{ and } 3 \geq x, x \in R \}.$$

Lesson 14: Sketching Square Roots Below and Above the Line y = x

The following introduces the graph of the function $f(x) = \sqrt{x}$ in relation to the diagonal line $y = x$ (dotted below).

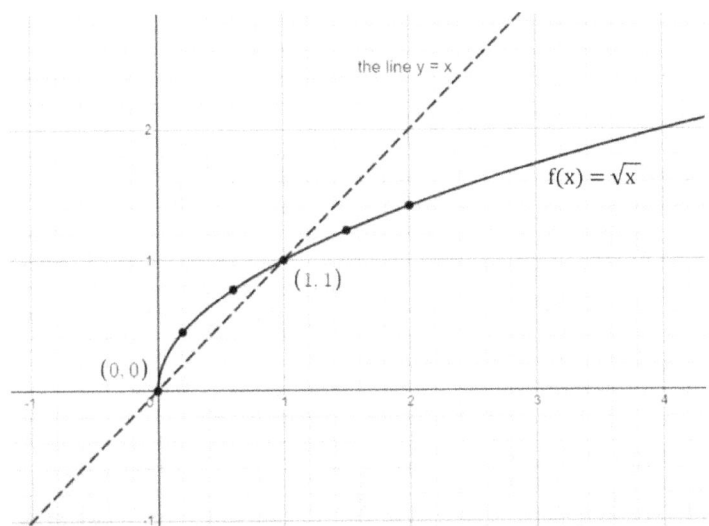

Figure 14.1

A well-known behavior of the square root function is examined in this section. The question here is whether the output y exceeds the value of the input x to the function $y = f(x) = \sqrt{x}$ in specific areas. The following clarifies.

The square root function produces values larger than their input x when the x is less than 1. When values of x are larger than x, then the output becomes always lower than the input used to calculate it. Examine *Figure 14.2* to better understand the statements:

when $x < 1$, y values are above the diagonal $y = x$ thus $y > x$,

when $x > 1$, y values are below the diagonal $y = x$ thus $y < x$,

when $x = 1$, then the y value is on the diagonal $y = x$ thus $y = x$.

Lex Sharp

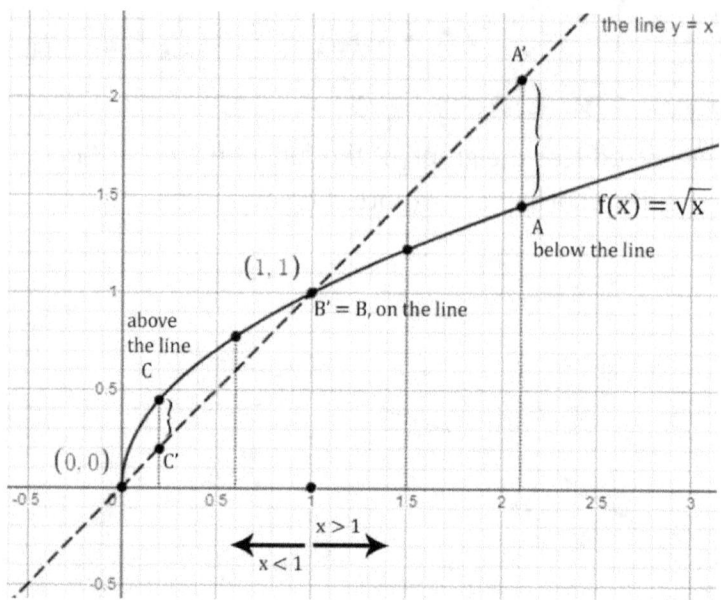

Figure 14.2

The x of point C in *Figure 14.2* is to the left of $x = 1$. The y value of C is above the dotted line: $C > C'$. All the y values in this area are above the dotted diagonal $y = x$.

The x of A is to the right of $x = 1$. The y value of A is below the dotted line: $A < A'$. All the y values in this area are below the dotted diagonal $y = x$.

The above/below switch in relation to the dotted line happens exactly at point $B(1, 1)$. This point is both on the line $y = x$ and the function $f(x)$.

This information is important when sketching the shape of the square root function. It is a good practice to mark the point $(1, 1)$ on the graph, and draw a line for the diagonal $y = x$ from $(0, 0)$ through and past $(1, 1)$. Then the curvature of the $\sqrt{x}$ plot travels:

> (i) above the diagonal in the area of $0 \leq x < 1$,
> (ii) meets (i.e. intercepts) the diagonal at $(1, 1)$, and
> (iii) ducks down under the diagonal for all $x > 1$.

Lesson 15: All Transformations

Generic Transformations Formula
The generic formula for function transformations is

$$g(x) = a * f(b(x - h)) + k.$$ (Expression 15.1)

Mapping Notation
The mapping notation equivalent to this formula is:

$$f \to g$$
$$(x, y) \to \left(\frac{1}{b}x + h, ay + k\right)$$ (Expression 15.2)

where:

- a and k are vertical only operations,
- b and h are horizontal only operations.

The coefficients are detailed in the next table.

Coefficient	Transformation Effect and Fully Qualified Name	Affected Coordinate/ Dimension
a, $a > 1$	represents a **vertical stretch** by a factor of a about the x-axis	y
b, $b > 1$	represents a **horizontal stretch** by a factor of $\frac{1}{b}$ about the y-axis	x
negative sign on a	represents a **vertical reflection** into the x-axis	y
negative sign on b	represents a **horizontal reflection** into the y-axis	x

h, in $(x - h)$	represents a **horizontal shift** by h units to - the Right if $h > 0$ - the Left if $h < 0$ h is identical to the h represented by the mapping (see *Expression 15.2*) and it is shown in the function formula as a subtraction (see *Expression 15.1*).	x
k	represents a **vertical shift** by k units - upwards if $k > 0$ - downwards if $k < 0$	y

Notice the following patterns in the use of language relating to these arrangements:

- Claiming a transformation is **into (or about) the x-axis**, is the same as saying it is a **vertical** transformation.
- If a transformation is said to be **vertical**, then it must be **into (or about) the x-axis**.
- Claiming a transformation is **into (or about) the y-axis**, is the same as saying it is a **horizontal** transformation.
- If a transformation is said to be **horizontal**, then it must be **into (or about) the y-axis**.
- A **vertical** transformation means changes to y.
- y is the vertical coordinate. If only changes to y occurred, the transformation is vertical and "into" or "about" the x-axis.
- A **horizontal** transformation means changes to x.
- x is the horizontal coordinate. If only changes to x occurred, the transformation is horizontal and "into" or "about" the y-axis.

Order of Transformations

Two transformations that apply to different dimensions can be made in any order. A horizontal and a vertical transformation does not interfere with one another. Each operates in its own dimension: either x only, or y only. It is customary to perform all stretches vertical and horizontal first, the following sequence marked by Roman numerals have identical outcomes.

- All horizontal transformations first, but so that the stretch (I) goes before the shift (II), followed by all vertical transformations so that the vertical stretch (III) is first then followed by the vertical shift (IV).

$$g(x) = af(b(x - h)) + k$$
$$\quad\quad\ \ (III)\ (I)\quad\quad (II)\quad (IV)$$

Desmos Link 15.1: https://sharpseries.ca/0/V1_L15.1.html

- All vertical transformations first with the stretch (I) happening before the vertical shift (II), followed by all horizontal transformations last, but so that the horizontal stretch (III) happens before the shift (IV).

$$g(x) = af(b(x - h)) + k$$
$$\quad\quad\ \ (I)\ (III)\quad\quad (IV)\quad (II)$$

Desmos Link 15.2: https://sharpseries.ca/0/V1_L15.2.html

- All stretches happen first beginning with the vertical (I) then the horizontal (II), or the other way around is also fine (II) then (I), since these happen in separate dimensions x and y. This is followed by all shifts in any order among themselves.

$$g(x) = af(b(x - h)) + k$$
$$\quad\quad\ \ (I)\ \ (II)\quad\quad (III)\quad (IV)$$
$$\quad\quad\ \ (II)\ (I)\quad\quad\ (IV)\quad (III)$$

Desmos Link 15.3: https://sharpseries.ca/0/V1_L15.3.html

Operations in the same dimension will interfere with each other. It matters if a stretch happens before a shift and vice-versa if they are in the same dimension, meaning either x only, or y only. The following statements clarify this:

- The order of *vertical* stretches and shifts matters and will interfere with each other in the same dimension: y. Performing the vertical shift before the vertical stretch is not the same transformation as having the vertical stretch first.
- The order of *horizontal* stretches and shifts matters and will interfere with each other in the same dimension: x. Performing the horizontal shift before the horizontal stretch is not the same transformation as performing the horizontal stretch first.

Horizontal Variations and Brackets in Function Formulas

As mentioned earlier, when multiple coefficients are used, the order of operations is important for the elements that are in the same dimension such as b and h below.

Attention must be paid to the function parameters in the formula

$$g(x) = a f(b(x - h)) + k.$$

i.e. the meaning of

$$f(b(x - h)), \quad \text{expression A,}$$

is quite different than the meaning without the use of brackets:

$$f(bx - h), \quad \text{expression B.}$$

Thus,

$$\text{expression A} \neq \text{expression B.}$$

Desmos Link 15.3: https://sharpseries.ca/0/V1_L15.4.html

The **use of brackets** in *expression A* represents **a stretch first order**. The lack of brackets in *expression B* means the shift occurred before the stretch.

If the intention is to represent a **stretch first order** (like *expression A*) and the function formula does not possess the brackets, then the formula must be rearranged via factoring. Brackets must be added to correctly represent the **horizontal stretch first** order. For example, if a horizontal stretch by a factor of $\frac{1}{3}$ is wanted before the horizontal shift in the function:

$$f(3x + 6),$$

then the function parameter must be rearranged to

$$f(\,3(x+2)\,), \quad \textit{according to expression A.}$$

With this constraint the horizontal shift becomes to the left by 2 units rather than 6.

However, without brackets

$$f(3x+6), \quad \textit{according to expression B,}$$

the formula implies the opposite order: a horizontal shift to the left by 6 units happening first, followed by a horizontal stretch by a factor $\frac{1}{3}$ at the end.

The number of units shifted in the two cases is different. The example according to *expression B* has a shift by 6 units while the earlier model according to *expression A*, shows a horizontal shift of only 2 units.

Formula vs. Mapping Notation, Correlation

The relationship between the coefficients of the function formula $a, b, h,$ and k in a transformation

$$g(x) = a\, f(\,b(x-h)\,) + k$$

and the factors of the **Mapping Notation** are shown in the table that follows.

Table 15.1

Formula Coefficient	Mapping Notation	Does the mapping factor equal the function formula coefficient?
a	$(x, y) \rightarrow (x, ay)$	Yes, the **vertical** stretch mapping factor is a.
b	$(x, y) \rightarrow (\frac{1}{b}x, y)$	No, the **horizonal** stretch mapping factor is the reciprocal of the formula coefficient, the factor is $\frac{1}{b}$.
$-h$	$(x, y) \rightarrow (x + h, y)$	No, the **horizontal** shift mapping has the opposite sign compared to the formula coefficient.
k	$(x, y) \rightarrow (x, y + k)$	Yes, the **vertical** shift mapping is identical to k.

The "**factor of a transformation**" is wording that frequently shows up in exam questions. It always refers to the constant of the **Mapping Notation** rather than the coefficients of the function formula.

It is especially important to focus the attention on the **horizontal** elements of the transformation when solving such questions since these have mapping notations that never equal the coefficients of the function formula.

When examining the behavior of a transformation, whether it is shifting left or right, stretching to thin or widen the graph, it is the **mapping** factors that provide the true visual effect of the movement and stretch trends:

- positive factors shift to the right,
- negative factors shift to the left,
- stretch factors between 0 and 1 act as diminishing percents and cause a thinning of the graph,
- factors above 1 cause a widening of the graph.

Notice in *Table 15.1* that the mapping factors are

- the **same** for **vertical** transformation as they are as the coefficients of function formulas,
- **different** mapping factors in **horizontal** transformation than the coefficients of the function formulas.

Lesson 16: Parameters Substitutions in Base Functions

One of the most common operations performed on a base function $f(x)$ is the handling of its parameter x for defining some transformation $g(x)$. This happens when the formula of a transformation $g(x)$ is calculated. Such formula shows $g(x)$ as a function of $f(x)$ rather than a function of x. This is best clarified by an example. The base function in this example is

$$f(x) = -x^2 + 2x,$$

and a transformation mapping is given below. A formula in x for the definition of $g(x)$ is required.

$$(x, y) \to \left(3x - 2,\ \tfrac{1}{2}y + 1\right),\quad \text{Expression 16.1}$$

To find the formula for $g(x)$, this example proceeds to examine the horizontal mapping first. The horizontal mapping is isolated here from the rest for simplicity:

$$(x,\ ) \to \left(3x - 2,\ \right),\quad \text{Expression 16.2}$$

It is the mapping that correlates to the visual aspect of the transformation and so the mapping expression $3x - 2$ simply follows the normal algebraic order of operations on the graph's coordinates: multiplication first and subtraction last. This means the stretch happens first because of the order of operations, and the shift happens last, i.e. the subtraction. This leads to a stretch-first as follows:

$$g(x) = a * f(\, b(x - h)\,) + k.$$

The given mapping showed a stretch by a factor of 3 first, which means the formula coefficient b equals the reciprocal of 3, that is

$\frac{1}{3}$. Since this stretch happens first, the shift that follows must appear in brackets (review *Lesson 15*, section "*Order of Transformations*").

The mapping states a horizontal shift to the Left by 2 units in *Expression 16.2*. Then the h coefficient inherited from the mapping must indicate a negative number for the Left direction, and in a moment, it will become clear that this becomes a positive number when processed into the $g(x)$ function formula:

$$h = -2.$$

Additionally, the function formula inherits this h with brackets, since the stretch happened before the shift:

$$b * (x - h) = b * (x - (-2)) = b * (x + 2).$$

Read more about horizontal variations and brackets in *Lesson 15*.

Next $b = \frac{1}{3}$ is also plugged in and the horizontal part of the expression is formed:

$$b * (x - h) = \frac{1}{3}(x + 2).$$

The vertical mapping is a lot simpler, a stretch by $\frac{1}{2}$ vertically and a shift upwards by 1 unit. Mapping and formula coefficients are the same in the vertical aspect. The overall equivalent function formula for the $g(x)$ becomes:

$$g(x) = \frac{1}{2} f\left(\frac{1}{3}(x + 2)\right) + 1 \quad \text{Expression 16.3}$$

So far, *Expression 16.3* can be seen as a variation on $f(x)$ rather than a variation on only x. In other words, $g(x)$ depends primarily on $f(x)$. For now, $g(x)$ "cannot see" past $f(x)$ so it cannot be defined in x only yet.

This is a fine definition, but better definitions rely on x only. Such was the definition of $f(x)$:

$$f(x) = -x^2 + 2x,$$

and $g(x)$ should be defined in a similar fashion. This can be done in two steps. The first step is figuring out the "inside" of the relationship to $f(\ldots)$, the second step is figuring out the "outside" of it as shown, the next steps clarify this:

1) the handling of the **inside** of $f(\ldots)$'s brackets in this example means plugging in the complicated parameter
$$\frac{1}{3}(x+2),$$
into $f(\ldots)$ and coming up with a result in x.

2) handling the **outside** of $f(\ldots)$ is about simply multiplying the result in Step 1 by $\frac{1}{2}$ and adding 1 to it.
$$\frac{1}{2}f(\ldots) + 1.$$

Step 1:

The question is how to plug in a complicated parameter to replace what is normally simply an x.

Instead of x, imagine a placeholder of "something else" more general in the definition of $f(\ldots)$. This placeholder can be replaced with anything just to show the pattern. For example, the hashed square glyph ▨ can play this role. Then ▨ replaces every occurrence of x. The original formula is shown on the left below and the pattern is contrasted on the right.

$$f(x) = -x^2 + 2x \qquad f(▨) = -▨^2 + 2▨$$

The glyph is only meant to helps think of $f(\ldots)$ in a more abstract way. The function can then be paraphrased so, to prove that nothing really changed:

Lessons, Math 30 Diploma Prep

The function $f(...)$ receives some unknown parameter ▨, this parameter is then squared and multiplied by -1. Next the parameter is doubled 2^*▨ and added the result.

The question becomes: what should be done if the parameter to $f(...)$ is no longer x but rather $\frac{1}{3}(x+2)$, like it is in *Expression 16.3*?

The pattern tracks this idea. It is known that

x and ▨ are interchangeable, and

▨ and $\frac{1}{3}(x+2)$ are interchangeable too.

Since ▨ is now representing a relatively complex expression, it's best to surround it with brackets. Therefore,

$f(▨) = -▨^2 + 2▨$ becomes $f(▨) = -(▨)^2 + 2(▨)$.

Next ▨ is replaced by the expression $\frac{1}{3}(x+2)$. The replacement must happen in every instance that ▨ occurs. This maintains the original pattern:

$$f(▨) = -(▨)^2 + 2(▨)$$
$$\downarrow \qquad \downarrow \qquad \downarrow$$
$$\tfrac{1}{3}(x+2) \quad \tfrac{1}{3}(x+2) \quad \tfrac{1}{3}(x+2)$$

The expression then equals

$$f\left(\tfrac{1}{3}(x+2)\right) = -\left(\tfrac{1}{3}(x+2)\right)^2 + 2\left(\tfrac{1}{3}(x+2)\right),$$

and after applying the square exponent and the opening of the brackets, the result is:

$$f\left(\tfrac{1}{3}(x+2)\right) =$$

$$-\left(\frac{1}{3}\right)^2 (x+2)^2 + 2\left(\frac{1}{3}(x+2)\right) =$$ The square is distributed into the brackets.

$$-\frac{1}{9}(x^2 + 4x + 4) + 2 * \frac{1}{3}(x+2) =$$ The redundant brackets were removed.

$$\frac{-1}{9} * (x^2 + 4x + 4) + \frac{2}{3}(x+2) =$$ Algebraic Rule 7 was applied.

$$-\frac{x^2 + 4x + 4}{9} + \frac{2(x+2)}{3} * 1 =$$ Identical denominators will be needed for the addition so that Algebraic Rule 6 can be applied.

$$-\frac{x^2 + 4x + 4}{9} + \frac{2(x+2)}{3} * \frac{3}{3} =$$ Algebraic Rule 7 is applied to multiply the fraction.

$$-\frac{x^2 + 4x + 4}{9} + \frac{2 * 3 * (x+2)}{9} =$$

$$\frac{-(x^2 + 4x + 4)}{9} + \frac{6 * (x+2)}{9} =$$ The two denominators are identical, the fractions can be added.

$$\frac{-x^2 - 4x - 4 + 6x + 12}{9} =$$ Like-terms together are bundled together.

$$\frac{-x^2 + 2x + 8}{9}$$

In conclusion:

$$f\left(\frac{1}{3}(x+2)\right) = \frac{-x^2 + 2x + 8}{9}.$$

This takes care of Step 1, i.e. the parameter in the brackets of *f(x)* has been resolved. This step represents only the horizontal operations that were part of the transformation *g(x)*. The vertical parameters are handled next.

Step 2:

The outside operations must be applied as $\frac{1}{2}f(...)+1$.

From there, the calculation is focused on replacing the results from Step 1 instead of $f(...)$. Then $f\left(\frac{1}{3}(x+2)\right)$ becomes a placeholder for the expression found in Step 1 as follows:

$g(x) = \frac{1}{2}f(...)+1 =$

$\frac{1}{2} * \frac{-x^2+2x+8}{9} + 1 =$ The expression found for $f(...)$ in Step 1 was plugged in.

$\frac{-x^2+2x+8}{18} + 1 =$ Algebraic Rule 7 must be applied.

$\frac{-x^2+2x+8}{18} + 1 * \frac{18}{18} =$ Identical denominators are necessary so the two fractions can be added.

$\frac{-x^2+2x+8+18}{18} =$ Like-terms are bundled together.

$\frac{-x^2+2x+26}{18}$

This concludes a proper function definition for $g(x)$ based on x only. All occurrences of $f(x)$ have completely disappeared:

$$g(x) = \frac{-x^2+2x+26}{18}.$$

Remembering the Process...

To remember this process, it is easy to think of the transition in Step 1 as a **"boxing the parameter"** metaphor and to visualize as shown next.

$$\text{boxing: } x \to \Box.$$

Then the transition in Step 2 is visualized as "**unboxing the parameter**" so it can become something else. This is a new more complex expression in every single occurrence of $\Box$:

$$\text{unboxing: } \Box \to \tfrac{1}{3}(x+2).$$

Lesson 17: In Depth Reflections

Reflections into axes are similar to real-life reflections into mirrors, although there are a few differences. In real-life the inside and outside of the mirror never mix, for transformations that is not always the case.

Two graphs are shown next: Figure 17.1 a reflection into the x-axis, and Figure 17.2 a reflection into the y-axis. The goal here is to generalize the understanding of reflections regardless of the axis used.

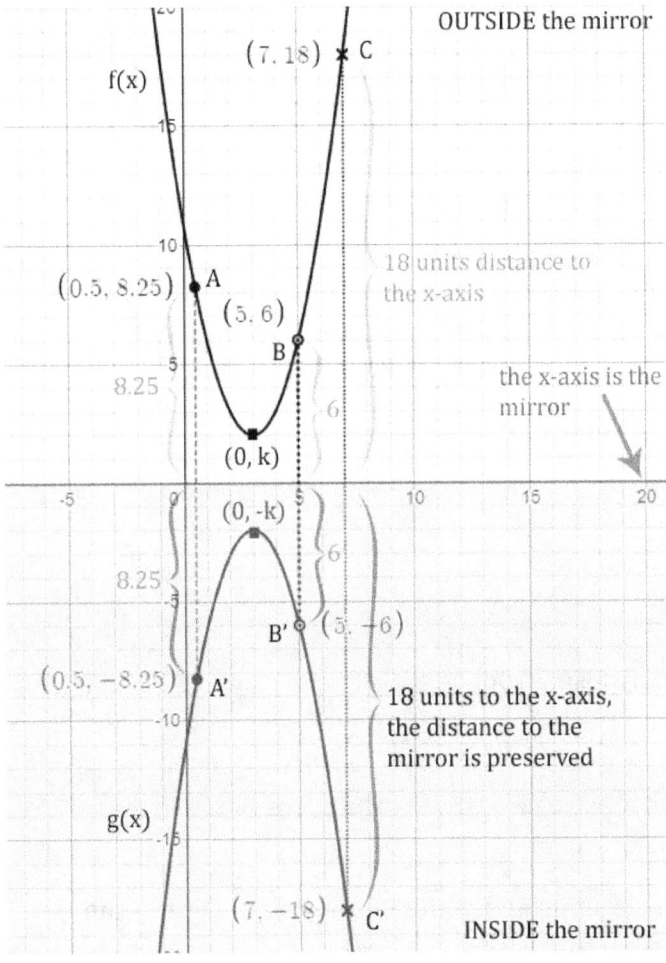

Figure 17.1, Desmos Link: https://sharpseries.ca/0/V1_L17.1.html

The graph in *Figure 17.1* depicts a reflection **into the x-axis**. The following attributes are applicable:

- The $f(x)$ shown was designed to fully reside "outside" the mirror. This will not always be the case with all functions.
- The mirror is seen as a line identical to the axis into which the reflection happens. In this example that is the x-axis.
- Since $f(x)$ is fully outside the mirror, symmetry demands that $g(x)$ will fully reside inside the mirror. The inside of the mirror is shown shaded at the bottom of the graph. This is everything below the x-axis for this example.
- All distances to the mirror (x-axis) are preserved. Distances to the x-axis are equivalent to y-values for each function. This is why reflections into the x-axis are seen as vertical, only the vertical coordinates y are affected.
- Elements residing literally on the mirror line are not affected and must remain unchanged. The $f(x)$ in Figure 17.1 didn't happen to have such points on the mirror, considering that the function resides above the axis (a later example will demonstrate such points).

Figure 17.1 shows point transitions:

$$A \to A',$$
$$B \to B',$$
$$C \to C',$$

the vertex is transposed from

$$(0, k) \to (0, -k).$$

Distances to the x-axis, a.k.a. the y-values, are preserved as follows:

- Point A is 8.25 units above the x-axis. Point A' is the same distance below the axis.

- Point B is 6 units above the x-axis. Point B' is the same distance below the axis.
- Point C is 18 units above the x-axis. Point C' is the same distance below the axis.

The above transformation stays with the same x values as the base function $f(x)$.

As a result, **the domain remains unchanged**.

The reflection only produces new y values on the opposite side of the x-axis. This means flipping the sign of y:

$$y \to -y.$$

If the base range is $\mathbf{R}$ (all the Real numbers) then it stays so for the transformation as well. However, if the base function has a range restriction, **the transformation range becomes its negative opposite**.

Because the reflection restricts itself to changes to the y-coordinates only, this makes the reflection vertical. The fully qualified name of such a reflection is: **vertical reflection into the x-axis**. The name depicts both the coordinate changed (vertical) and the name of the mirror used (the x-axis).

With the quadratic function shown in this example there is a minimum $y = k$ causing the base range to be

$$y \geq k \quad \text{Expression 17.1.}$$

The reflection turns this into **the opposite range**:

$$y \leq -k \quad \text{Expression 17.2.}$$

This makes sense considering **the mapping** of reflections is equivalent to a vertical stretch by a factor of -1:

$$(x, y) \to (x, -y).$$

The conclusion is the **transformation function formula** is:

$$g(x) = -f(x).$$

The graph in *Figure 17.2* shows another reflection, this time **into the y-axis**. The following attributes are applicable:

- The $f(x)$ resides mostly outside the mirror, and slightly infringes on the inside of the mirror. Such infringements are acceptable.
- The mirror is still seen as a line identical to the axis into which the reflection happens. In this example the mirror is the y-axis and its inside appears shaded on the left of the graph.
- Since $f(x)$ does not fully reside on the outside of the mirror, symmetry demands that neither does $g(x)$ fully reside on its own side.
- All distances to the mirror (y-axis) are preserved. Distances to the mirror are equivalent to the x-values of each function. This is why such a reflection is known to be horizontal, it only affects the horizontal coordinate x.
- All elements residing literally on the line that is the mirror, are not affected and must remain the same. This $f(x)$ has a y-intercept which is positioned exactly on top of the y-axis. This point becomes **invariant** to the transformation since it must not be affected. Another way to think of it is to see that its distance away from the mirror is zero, since it resides on top. The transformation preserves this distance which is why it remains 0 and thus on top of the line.

The range of this reflection is not affected, the same y-values are applicable as for the base function $f(x)$. This is because this reflection does not operate at all on the y coordinate.

If the base domain is $\mathbb{R}$ (all the Real numbers) then it stays so for the transformation as well, and indeed that is the case for this example.

When the base function has a domain restriction, **the**

transformation domain becomes its negative opposite. This example has no domain restrictions.

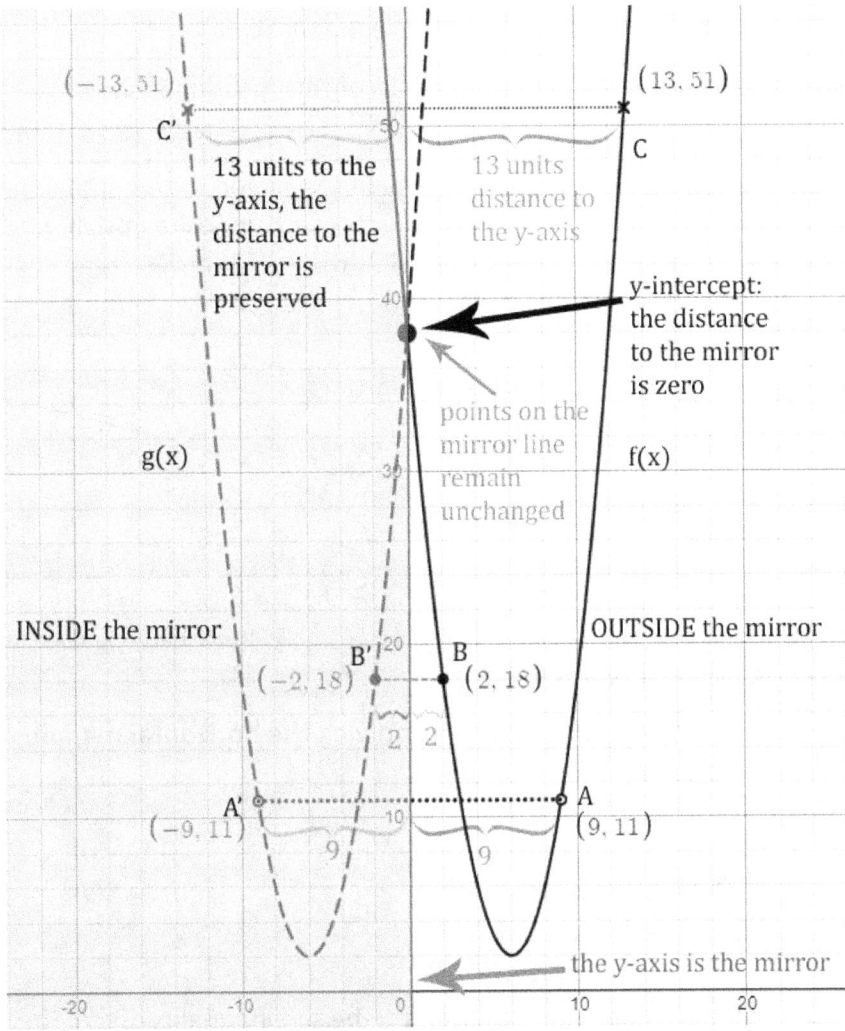

Figure 17.2, Desmos Link: https://sharpseries.ca/0/V1_L17.2.html

Examine the annotations on *Figure 17.2* for various details.

The next example illustrates a function with a domain restriction and shows what happens to it after a horizontal reflection into the y-axis is applied:

$$f(x) = \sqrt{x-1}.$$

This function uses a square root. Its domain is $x - 1 \geq 0$. Read more about domain restrictions in *Lesson 13*.

After solving for x the solution to the inequality is $x \geq 1$.

A horizontal reflection into the y-axis of this function is shown in *Figure 17.3*.

The reflection flips the original domain restriction to the other side of the mirror (y-axis), becoming its exact opposite in sign, i.e. $x \leq -1$ while preserving the distance of the domain from the "surface" of the mirror, i.e. the y-axis.

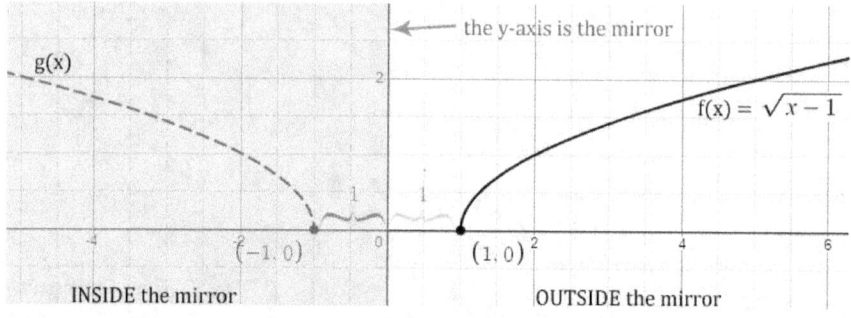

Figure 17.3, Desmos Link: https://sharpseries.ca/0/V1_L17.3.html

The new domain is calculated by applying **the horizontal mapping** to the coordinates:

$$(x, y) \rightarrow (-x, y).$$

The transformation formula is:

$$g(x) = f(-x).$$

The original $f(x)$ domain was found to be the inequality $x \geq 1$. The mapping factor is a horizontal stretch by a factor of (-1). Then the parameter $-x$ is to replace x as follows

$$g(x) = f(-x) = \sqrt{\boxed{} - 1} = \sqrt{-x - 1}.$$

To help with this transition the boxing/unboxing pattern from *Lesson 16* can be used by plugging in $-x$ instead of x, i.e.:

boxing: $x \rightarrow \boxed{}$,

unboxing: $\boxed{} \rightarrow (-x)$.

Since $g(x) = \sqrt{-x-1}$ the new domain depends on the inequality

$$-x - 1 \geq 0.$$

Solving this inequality is identical to the process of isolating x. After adding -1 to both sides of the inequality, the result is getting closer to an isolated x:

$$-x - 1 + 1 \geq 0 + 1,$$

$$-x \geq 1.$$

This doesn't yet fully isolate x, but it rather isolates $-x$.

Multiplying by -1 removes the negative sign from x, but it also flips the direction of the inequality like so:

$$-x * (-1) \leq (-1) * 1,$$

$$x \leq -1.$$

Thus the transformation's domain becomes $x \leq -1$. This shows what is meant by a **"flipping of the domain"**: if the original domain was x values larger than 1, then 1 is seen here as a type of a borderline. The borderline 1 is flipped to -1 and the values of x are now smaller than the borderline, rather than larger, thus the flipped domain is $x \leq -1$.

Summary:

Reflections into the **x-axis**	Reflections into the **y-axis**
In both cases, "reflecting into" infers there is a mirror of the reflection.	
The mirror = **x-axis**	The mirror = **y-axis**

"Into the x-axis" always means a **vertical** transformation.	"Into the y-axis" always means a **horizontal** transformation
The **domain** is unchanged by the transformation.	The **range** is unchanged by the transformation.
If the **base range** is $\mathbb{R}$ then it stays so for the transformation as well.	If the **base domain** is $\mathbb{R}$ then it stays so for the transformation as well.
If the **base range is an inequality**, i.e. has a restriction, then it must be flipped by *-1* i.e. transformed y-values get a **new range**: $y \geq k \to y \leq -k$, or $y \leq k \to y \geq -k$. Minimum values become maximum values. Maximum values become minimum values. All points and the **range** of the base function are flipped to the opposite side of the mirror by the reflection.	If the **base domain is an inequality**, i.e. has a restriction, then it must be flipped by *-1*, i.e. transformed x-values get a **new domain**: $x \geq p \to x \leq -p$, or $x \leq p \to x \geq -p$. Leftmost values become rightmost values. Rightmost values become leftmost values. All points points and the **domain** of the base function are flipped to the opposite side of the mirror by the reflection.
In both cases, points **on the axis** that represents the surface mirror of the reflection are **invariant**, and thus unaffected by the reflection. These remain distanced *0* units away from the axis.	
x-intercepts, i.e. the roots reside on the mirror = x-axis,	y-intercepts reside on the mirror = y-axis, thus they are

thus they are invariant.	invariant.
All distances to the mirror = x-axis, are preserved on the other side of the mirror. These are the y-values of the transformation.	All distances to the mirror = y-axis, are preserved on the other side of the mirror. These are the x-values of the transformation.

In both cases, base functions and their reflections are completely symmetric in relation to the axis representing the mirror.

Tracking the words in the two columns: range, domain, into x, into y, vertical, horizontal, x-axis, y-axis reveals a pattern by which they consistently correlate across the two columns to their exact opposites:

$$\text{Domain} \leftrightarrow \text{Range}$$

$$\text{x-axis} \leftrightarrow \text{y-axis}$$

$$\text{Into the x-axis} \leftrightarrow \text{Into the y-axis}$$

$$\text{Horizontal} \leftrightarrow \text{Vertical}$$

Note that inverse functions reflect into the diagonal $y = x$ rather than into one of the x or y axes. The rules listed above can be slightly adapted to reflect inverse functions as well but are not represented in the table above.

Lesson 18: Generating an Inverse

The method for generating an inverse $g(x)$ for a base function $f(x)$ involves a few steps:

1. the swapping of the x and the y in the definition of $f(x)$,
2. the isolation of y, which is equivalent to solving for y, in the expression accomplished in Step 1. This y becomes the inverse function $y = g(x)$.
3. $g(x)$ may, or may not be a function, could end up being simply a relation.

This is better understood with an example.

Example 1

Assume the base function is $y = f(x) = x^2 - 6x + 9$, and the name given to the inverse is going to be $g(x)$. The calculation of $g(x)$ is as follows.

Step 1:

The base function is: $y = x^2 - 6x + 9$. All occurrences of x are changed to y and vice-versa. This swap leads to the relation:

$$x = y^2 - 6y + 9.$$

Step 2:

The expression gained from Step 1 is used to solve for y. This y is the inverse $y = g(x)$. Solving for y is identical to isolating y from the expression

$$x = y^2 - 6y + 9.$$

The difficulty in doing so derives from the fact this expression is carrying

1. a square exponent y^2 as well as
2. a linear term y.

This causes the same problems as having two separate variables to solve from a single equation rather than having only one variable. This is temporarily remedied by introducing one binomial to stand in for a single variable, of the form

$$y^2 - 6y + 9 = (y - number)^2$$

to solve that problem. Temporarily, the equation is solved for this binomial instead of y. Solving for the binomial is the same as isolating it as shown next. Then the concepts in *Lesson 2* are used to get rid of the square exponent which will then permit a return to solving for y.

The first goal is to find the binomial $(y - number)^2$ that is equivalent to $y^2 - 6y + 9$. The expression $y^2 - 6y + 9$ can be either factored or *Algebraic Rule 11* can be applied to recognize it a perfect square, i.e.:

$$y^2 - 6y + 9 =$$

$$y^2 - 2 * 3 * y + 3^2 =$$

$$(y - 3)^2.$$

The equation is now based on this binomial, but the plan remains the same: to eventually solve for y

$$x = (y - 3)^2.$$

Since this is an equality the two sides can swapped. y is wanted on the left side so that the expression gets closer to something like $y = \ldots$ which is how y becomes isolated.

$$(y - 3)^2 = x.$$

For now, this move has isolated the binomial to the left side. It helps that no y values appear on the right anymore. An even closer form to an isolated y can be achieved by isolating $y - 3$ instead of its squared version. This is easily done by applying a square root to both sides of the equality. It will have to be accompanied by adding $\pm$ as described in *Lesson 2*, since the origin

63

of the expression is a quadratic: $(y - 3)^2$, therefore

$$y - 3 = \pm \sqrt{x}.$$

Still, y does not stand alone isolated on the left yet. After adding the number 3 to both sides of the equation:

$$y - 3 + 3 = \pm \sqrt{x} + 3,$$

$$y = \pm \sqrt{x} + 3$$

y is finally completed isolated on its own on the left, and thus the equation is solved for y.

This completes Step 2, so the inverse is

$$y = g(x) = \pm \sqrt{x} + 3.$$

An alternate way to represent $g(x)$ without the $\pm$ signs, is to say that $g(x)$ is the union of two separate functions:

$$g_1(x) = \sqrt{x} + 3, \text{ and}$$

$$g_2(x) = -\sqrt{x} + 3.$$

The plotting of $g(x)$ requires the plotting of both $g_1(x)$ and $g_2(x)$ on the same graph.

Additional Comments:

It can be clear from the use of $\pm$ that $g(x)$ is producing two y values for every input x. One is positive and the other negative. The conclusion from this is that $g(x)$ is not a function, but simply a formula. The $g(x)$ function would not pass the vertical line test yielding two separate y values for each x in the domain.

When functions that prefix their output with $\pm$ are seen, the quick conclusion is these are only relations, not functions. This is common for the inverses of quadratic functions that have no domain restrictions.

The example used above was made easy by ensuring the expression in y was a perfect square: $y^2 - 6y + 9 = (y - 3)^2$. The next example demonstrates a case without a perfect square. The **Complete the Square** method is used instead as detailed.

Example 2

A base function $y = f(x) = x^2 - 6x + 8$ is assumed. The name given to the inverse is $g(x)$ like before. The calculation of $g(x)$ follows.

Step 1:

The base function is $y = x^2 - 6x + 8$. All occurrences of x are changed to y and vice-versa. This leads to

$$x = y^2 - 6y + 8.$$

Step 2:

The expression gained from Step 1 must be used to solve for y, and $y = g(x) = \textit{the inverse}$.

Solving for y is identical to isolating y from $x = y^2 - 6y + 8$.

For this it would be advantageous to get a binomial in y that is a perfect square as seen in the previous example. However, the expression $y^2 - 6y + 8$ is not a perfect square.

The next best thing is a **perfect square plus a constant imperfection**, such expression is available for all quadratics that are not perfect squares. The expression can be accomplished via the **Complete the Square** method described in *Lesson 5*. When the method is applied the result is:

$$y^2 - 6y + 8 =$$

$$y^2 - 2 * \frac{6}{2} * y + 8.$$

Then the following zero $0 = \left(\frac{6}{2}\right)^2 - \left(\frac{6}{2}\right)^2 = 3^2 - 3^2$ is added to the expression. Rearranging with such a zero does not disturb the original value of the expression since adding a zero has no effect. The method continues to rearrange the terms into the perfect square plus a constant imperfection as shown next.

$$y^2 - 2*\frac{6}{2}y + 0 + 8 =$$

$$y^2 - 2*\frac{6}{2}y + 3^2 - 3^2 + 8 =$$

$$y^2 - 2*3y + 3^2 - 3^2 + 8$$

The regrouping of terms based on a perfect square is shown in brackets below. Whatever is leftover it contributes to the constant that equals the imperfection.

$$(y^2 - 2*3y + 3^2) - 3^2 + 8$$

The terms in brackets are spelled out as a perfect square. The constant terms of the imperfection are bundled together into a single number.

$$(y-3)^2 - 9 + 8 =$$

$$(y-3)^2 - 1$$

The above gives a "complete the square" description of the expression $y^2 - 6y + 8$. This produces the binomial discussed earlier which is temporarily solved for. The original expression is now equivalent to:

$$x = y^2 - 6y + 8,$$

$$x = (y-3)^2 - 1.$$

It's an equality so it's OK to switch sides. The variable y is wanted on the left.

$$(y-3)^2 - 1 = x.$$

Lessons, Math 30 Diploma Prep

First, the binomial in y is isolated. The first step in doing so is to remove the number -1 from the left. This is done by adding 1 to both sides of the equation.

$$(y-3)^2 - 1 + 1 = x + 1,$$

$$(y-3)^2 = x + 1.$$

The squared expression $(y-3)^2$ is closer to y than $(y-3)^2 - 1$ was, a small progress occurred. The square exponent interferes with access to isolating y, so it would be good to remove. From here getting rid of the square exponent is easy based on *Lesson 2*, a squaring of both sides of the equation is used:

$$y - 3 = \pm \sqrt{x+1}.$$

Yet again, $y - 3$ is closer to isolating y than $(y-3)^2$ was. Since -3 interferes with the isolating of y, it is added to both sides of the equation which has the effect of removing it.

$$y - 3 + 3 = \pm \sqrt{x+1} + 3,$$

$$y + 0 = \pm \sqrt{x+1} + 3,$$

$$y = \pm \sqrt{x+1} + 3.$$

This concludes the calculation of y, equivalent to $g(x)$ as the inverse of $f(x)$:

$$y = g(x) = \pm \sqrt{x+1} + 3.$$

Again, an alternate way to represent $g(x)$ is to say it is the union of two separate functions: $g_1(x)$ and $g_2(x)$, one for each sign:

$$g_1(x) = \sqrt{x+1} + 3, \text{ and}$$

$$g_2(x) = -\sqrt{x+1} + 3.$$

The plotting of $g(x)$ requires the plotting of both $g_1(x)$ and $g_2(x)$ on the same graph. The formula for $g(x)$ does not represent a function due to it not passing the vertical line test. It produces two y values (one positive and one negative) for each value of x.

In Summary for a General Approach:

After swapping the x and y of the base function, the inverse of a parabola is derived from a squared binomial in y. The binomial appears either

(i) by itself as a perfect square, or
(ii) a perfect square accompanied by an imperfection that is a constant.

In both cases the squared binomial is isolated on the left and then both sides are square-rooted and accompanied by a $\pm$ to find the inverse.

Lesson 19: Factoring Quadratics Based on Sum and Product of Roots

The function $y = x^2 + bx + c$ can be rearranged to represent the following model:

$$y = x^2 - sx + p.$$

This is a special case of the quadratic because the coefficient of x^2 is 1, rather than some generic constant a.

Initially there are two roots assumed: x_1 and x_2. These roots could be different or equal. If the roots do not exist, then the quadratic cannot be factored, in which case none of this is applicable.

The letter s stands for "sum" and is:

$$s = x_1 + x_2.$$

The letter p stands for "product" and is:

$$p = x_1 * x_2.$$

If the (real) roots exist, then the function is equal to the factoring form:

$$y = x^2 - sx + p = (x - x_1)(x - x_2).$$

The sum coefficient s is accompanied by a minus. This method works best when it is easy to rearrange a quadratic in a way that permits noticing which two numbers compound into s for sum and p for product. For example, this function is given:

$$y = x^2 + 5x + 6.$$

After formatting the sum coefficient s with a minus the correct form is:

$$y = x^2 - (-5)x + 6$$

The two numbers that have a sum of -5 and a product of 6 are: -3 and -2:

$$x_1 = -3,$$

$$x_2 = -2.$$

Therefore the factoring is derived as:

$$y = x^2 + 5x + 6 = (x - x_1)(x - x_2) =$$

$$(x - (-2))\,(x - (-3)) =$$

$$(x + 2)(x + 3)$$

which is the factored form based on the sum and the product.

The equation was given in the form $y = x^2 - sx + p$, which is very close to standard form. The leading coefficient of 1. Things are a bit more complicated when the coefficient is not 1, but typically the same can be accomplished by factoring out the leading coefficient first. For example the terms in

$$y = 2x^2 - 8x + 6$$

are all divisible by 2. Thus 2 can be factored out in front of brackets:

$$y = 2 * x^2 - 2 * 4x + 2 * 3$$

$$y = 2(x^2 - 4x + 3)$$

Then the full expression can be set aside for a moment and instead, only the inner brackets expression is treated to the previous method. Thus,

$$x^2 - sx + p = (x - x_1)(x - x_2) = x^2 - 4x + 3,$$

is analyzed first. Two numbers x_1 and x_2 can be guessed for this expression to fit the sum and product model:

$$s = 4 = 3 + 1, \text{ and}$$

$$p = 3 = 3 * 1.$$

Therefore $x_1 = 3$ and $x_2 = 1$.

The expression in the brackets is factored

$$x^2 - 4x + 3 = (x - 3)(x - 1).$$

Restoring the full expression is accomplished by substituting the two factors as shown:

$$y = 2(x^2 - 4x + 3),$$

$$y = 2(x - 3)(x - 1).$$

Lesson 20: Inequalities with Functions

Any quadratic function

$$y = f(x) = ax^2 + bx + c$$

can be plugged into an inequality to produce y values that are smaller or larger than some specific number. These inequalities look like this typically

$$\text{some expression in } x \geq a$$

which is the same as saying

$$y = f(x) \geq a.$$

Here a can be any number, positive, or negative, and any one of the inequality symbols can be used from the set:

$$\geq, >, \leq, <.$$

Here are a few examples of inequalities:

$$2x^2 - 8x + 8 < 0,$$

$$3x - 6 \geq 8,$$

$$-3x^2 + 6x - 3 \leq -2,$$

$$-(x-2)^3 > -27.$$

In essence, whenever an inequality can be rearranged as $f(x) > a$, (or with some other inequality symbol) the template to solve consists of restating the inequality as a question, or to paraphrase the mathematical expression as:

> "What are the values of x, so that y values can be produced that are $y > a$, so that $y = f(x)$?"

The question helps visualize the problem and is always adapted to the inequality symbol used and whatever the number a might be.

Such inequality can be solved graphically by verifying where the line $y = a$ intersects the graph of $f(x)$ and which are its y values that position correctly. For example, this is above the line $y = a$ if

$$y = f(x) > a,$$

or below the line $y = a$ if

$$y = f(x) < a.$$

Frequently an inequality such as $y = f(x) > a$ is rearranged so that the constant a is subtracted from both sides:

$$f(x) - a > a - a, \text{ then}$$

$$g(x) > 0, \text{ where } g(x) = f(x) - a.$$

In the end, the inequality $g(x) > 0$ is solved for x instead of solving the original $f(x) > a$. Then instead of relating to the line $y = a$, the line $y = 0$ can be taken, which represents the x-axis.

Example 1

The inequality $x^2 \geq 0$, must be solved for x. Before solving, the inequality can be understood by paraphrasing the question as noted before. That is,

> "What are the values of x, so that y values can be produced that are $y \geq 0$, so that $y = f(x) = x^2$?"

This is a very simple inequality because the range of this function is already precisely as needed by the question. That is, the function produces only values larger than or equal to 0. This is because the square flips all expressions into positive values. Thus all the values $y = f(x) \geq 0$ behave so by default. This behavior does not depend on specific x values, all x values in the entire range $\mathbb{R}$ are solutions to this inequality. This is also obvious when the graph is examined visually.

The visual aspect is where the above paraphrasing helps to understand what to look for. The function is $y = x^2$. It has two identical roots $x_1 = x_2 = 0$. Because the roots are identical, this graph touches the x-axis in a single point (is tangent) as shown in *Figure 20.1*. Because of this the function only stays on one side of the x-axis, in this case above, and will not go past it.

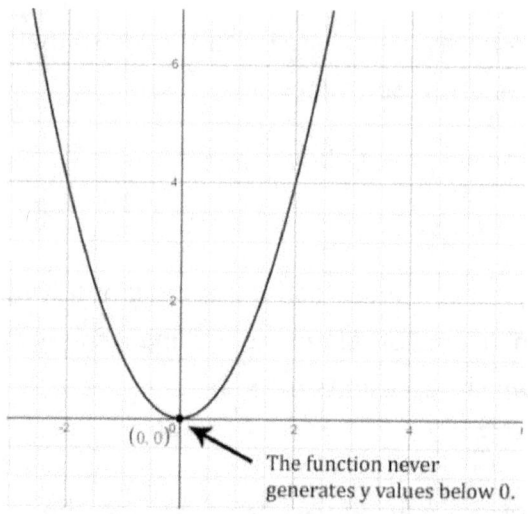

The arms of x^2 point upwards because the leading coefficient is *1* and thus positive (see *Lesson 8*). There is no doubt the range, i.e. the y values stay above or at 0.

Figure 20.1

The solution to the inequality is given as an expression of x. It can be formalized as

$$\{ x \mid x \in R \},$$ i.e. all the Real numbers.

The next example explores a parabola that has two distinct x-intercepts $x_1 \neq x_2$ rather than just one. The inequality is going to be of the form

$$f(x) > 0, \text{ or}$$
$$f(x) < 0,$$

and must be solved for x.

Example 2

If $f(x) = x^2 - 9$ is used, factoring can provide the roots.

The factoring in this case is resolved based on a "difference of squares" (see *Algebraic Rule 8*), because both x^2 and 9 are perfect squares.

$$y = f(x) = x^2 - 9 =$$

$$x^2 - 3^2 =$$

$$(x - 3)(x + 3),$$

then the roots are

$$x_1 = 3, \text{ and } x_2 = -3.$$

The parabola

$$y = f(x) = x^2 - 9 = (x - 3)(x + 3)$$

is very easy to sketch. Then the y values' relation to 0 (a.k.a. the x-axis or $y = 0$) can be observed on the graph rather than analyzing it algebraically.

The leading term of this function is: x^2. When writing the term as $1 * x^2$ the leading coefficient 1 is identified. This is a positive number. Then the arms of this parabola point upwards, and the y-intercept is the constant term of the function formula in standard form, i.e. (-9). Read more about y-intercepts in *Lesson 9*.

Two vertical lines are drawn through the two x-intercepts and are shown in *Figure 20.2*. All x values between the two vertical lines are naturally referred to as **"between the roots"** and are shown in the unshaded area of the graph. Two additional intervals form as well on the left and the right side of these vertical lines. These two areas together are referred to as **"outside the roots"** and are shown the two shaded areas in *Figure 20.2*.

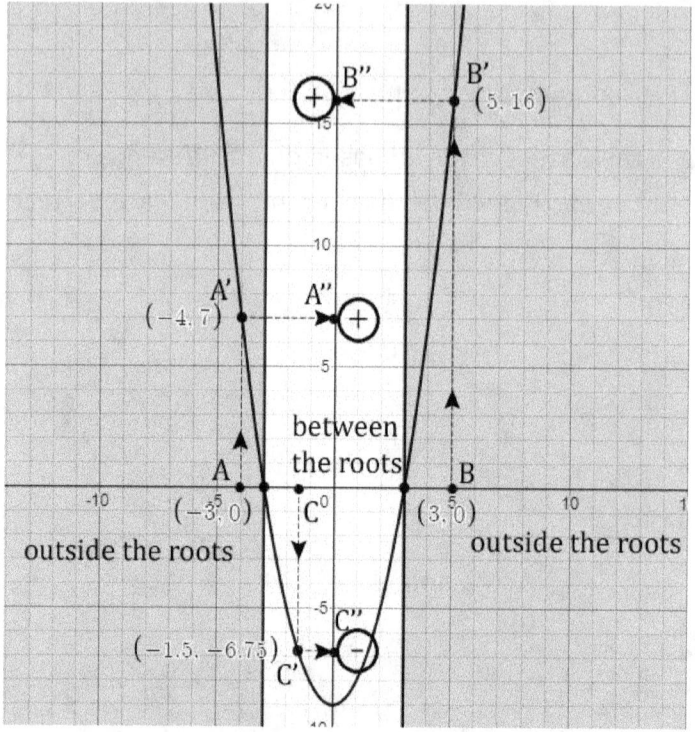

Figure 20.2, Desmos Link: https://sharpseries.ca/0/V1_L20.2.html

All the x values between the roots produce negative y values. Positive y-values are found for x values outside the roots. To better understand this idea, points are tracked on the graph, click on the Desmos link to see an animation.

For example, the x coordinate of point C designates a position "between the roots". Point C relates to point C' which resides on the plot. Furthermore C' relates to its y value on the y-axis leading to point C''. The relationships between these points are described below in three equivalent statements:

$$\text{from the x axis} \rightarrow \text{to the plot of } f(x) \rightarrow \text{to the y axis:}$$

$$C \rightarrow C' \rightarrow C''$$

$$\text{from an x between the roots} \rightarrow \text{to the plot of } f(x) \rightarrow \text{to negative y.}$$

The points A and B in Figure 20.2 are found **"outside the roots"**

Lessons, Math 30 Diploma Prep

and are examined in a similar fashion to point C:

from the x axis → to the plot of f(x) → to the y axis

$$A \to A' \to A'', \text{ and}$$

$$B \to B' \to B'',$$

from an x outside the roots → to the plot of f(x) → positive y.

Therefore, for any parabola with the **arms up** (a.k.a. Smiley Parabola – see *Lesson 8*):

- x values **between** the roots lead to **negative** y values,
- x values **outside** the roots lead to **positive** y values.

Example 3

The function $f(x) = -x^2 + 9$ is given. Here the arms of the parabola point downwards. This is so because the leading coefficient -1 is negative. It is part of the leading term $-x^2$. The parabola is shown in *Figure 20.3*. The following is applicable:

- x values **between** the roots lead to **positive** y values,
- x values **outside** the roots lead to **negative** y values.

In this case the point correlation is described by the following statements and can be tracked in *Figure 20.3*:

from the x axis → to the plot of f(x) → to the y axis

$$C \to C' \to C''$$

from an x between the roots → the plot of f(x) → positive y values.

Points A and B are "**outside the roots**":

$$A \to A' \to A'', \text{ and}$$

$$B \to B' \to B'',$$

from an x outside the roots → the plot of f(x) → negative y values.

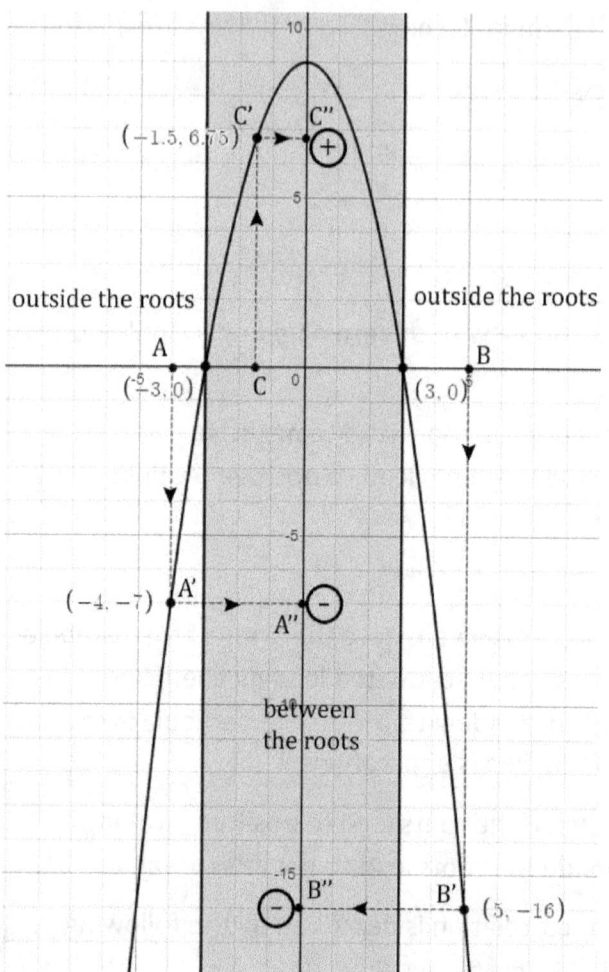

Figure 20.3, Desmos Link: https://sharpseries.ca/0/V1_L20.3.html

In both **Example 2** and **Example 3**, with the arms of the parabola up or down, the graphs show how y values correlate to their x counterparts as intervals. These intervals are translated from the inside/outside of the roots, which solve quadratic inequalities for x in relation to the $y = 0$, a.k.a. the x-axis. The following helps clarify this.

The function $f(x) = (x - 3)(x + 3)$ shown in *Figure 20.2* for the inequality $f(x) \geq 0$, has a solution of x values that are outside the roots to ensure positive or zero y-s. The solution translates into the two shaded intervals of *Figure 20.2*:

$$x \leq -3 \text{ and } 3 \geq x.$$

In *Figure 20.3* a mirror image of the previous function is depicted: $y = -f(x) \geq 0$ and the solution is the opposite: x values between the roots, i.e. the solution in x to get positive or zero y values is

$$-3 \leq x \leq +3.$$

In Conclusion:

The solution for Example 1 in Figure 20.2 for the inequality

$$y = f(x) = (x-3)(x+3) \geq 0 \text{ is}$$

$$x \leq -3 \text{ and } 3 \geq x.$$

The solution for Example 2 in Figure 20.3 for the inequality

$$y = f(x) = -(x-3)(x+3) \geq 0 \text{ is}$$

$$-3 \leq x \leq +3.$$

Typical Inequalities

In many cases, it is the domain of a function that needs to be solved as an inequality. For example, when a function $f(x) = x^2 - 9 = (x-3)(x+3)$ is used in a square root such as

$$g(x) = \sqrt{f(x)} = \sqrt{(x-3)(x+3)},$$

the domain of $g(x)$ must at all times be positive or zero for the square root to be defined. This means that only values of x that produce positive or zero y values in the parabola given by $f(x)$ will be permitted.

The domain of $g(x)$ is the same as the solution to the inequality $y = f(x) \geq 0$. This specific example was solved above (see Example 2) to be $x \leq -3$ and $3 \geq x$. Thus this is where $g(x)$, which is identical to where the square root of $f(x)$ is defined.

Techniques Used in Solving Inequalities

Earlier, it was noted that techniques similar to solving equations were applied to inequalities. Such was the subtraction of the constant a in:

$$f(x) > a,$$
$$f(x) - a > a - a,$$
$$f(x) - a > 0.$$

The subtraction can be applied as long as this is done to both sides of the inequality. Additions are also allowed in a similar manner:

$$f(x) > -b,$$
$$f(x) + b > -b + b,$$
$$f(x) + b > 0.$$

Multiplication and division are allowed. However, multiplications and divisions are bound by a caveat: the inequality symbol must be flipped to its opposite whenever a negative number is used to multiply or divide the two sides.

Consider c to be a positive number $c > 0$. Because it is positive, the above rule does not apply, and the inequality unfolds in the typical way for such c for example when used in the following inequality:

$$\frac{1}{c} f(x) > 1$$

The positive constant c can be multiplied on both sides of the inequality with no concerns like so:

$$\frac{1}{c} * c * f(x) > 1 * c,$$
$$\frac{1}{\cancel{c}} * \cancel{c} * f(x) > c,$$

$$1 * f(x) > c,$$

$$f(x) > c.$$

However, if instead a negative constant $c < 0$ value is used, then the multiplication has the added effect of **flipping the symbol of the inequality** from

(i) $<$ to $>$ and vice-versa, and from
(ii) $\leq$ to $\geq$ and vice-versa.

Then using the same inequality as before the progress looks like so:

$$\frac{1}{c} f(x) > 1,$$

$$\frac{1}{c} * f(x) * c < 1 * c, \quad (if\ c < 0)$$

$$\frac{1}{\cancel{c}} * \cancel{c} * f(x) < 1 * c,$$

$$f(x) < c.$$

In conclusion the solution for

$$\frac{1}{c} * f(x) > 1 \text{ is:}$$

$f(x) > c$ when c is positive,

$f(x) < c$ when c is negative.

Example 4

Solve the inequality $-2x + 4 > 0$ for x. Provide an algebraic solution.

Solution

Just like in equations, solving for x is the same as isolating x to one side of the inequality. Initially there are two things that stand in the way of isolating x:

(i) its multiplication by (-2), and
(ii) its addition by 4.

These need to be removed in the following order: the addition is removed first, then the multiplication.

$-2x + 4 > 0$

$-2x + 4 - 4 > -4$

$-2x + 0 > -4$

The goal is to remove 4 from the left. For this 4 is subtracted from **both** sides of the inequality. This begins the effort to isolate x.

$-2x > -4$

$(-1) * 2 * x > -4$

x is accompanied by two multiplications in its $-2x$ form:

- multiplication by (-1) and
- multiplication by 2.

These will be removed one by one.

$(-1) * 2 * x * (-1) < -4 * (-1)$

$(-1) * (-1) * 2 * x * < +4$

Multiplication by (-1) on both sides of the inequality has the role of removing (-1), a.k.a. the negative sign, from the left side where x is currently being isolated. The negative number flipped the inequality symbol from larger to smaller.

$2 * x < 4$

$\frac{2}{2} * x < \frac{4}{2}$

Both sides are divided by 2 to isolate x. There is no need to worry about the $<$ symbol from here because 2 is a positive number.

$x < 2$ This provides the algebraic solution for the original inequality $-2x + 4 > 0$.

Solving the $-2x + 4 > 0$ inequality graphically, looks like so:

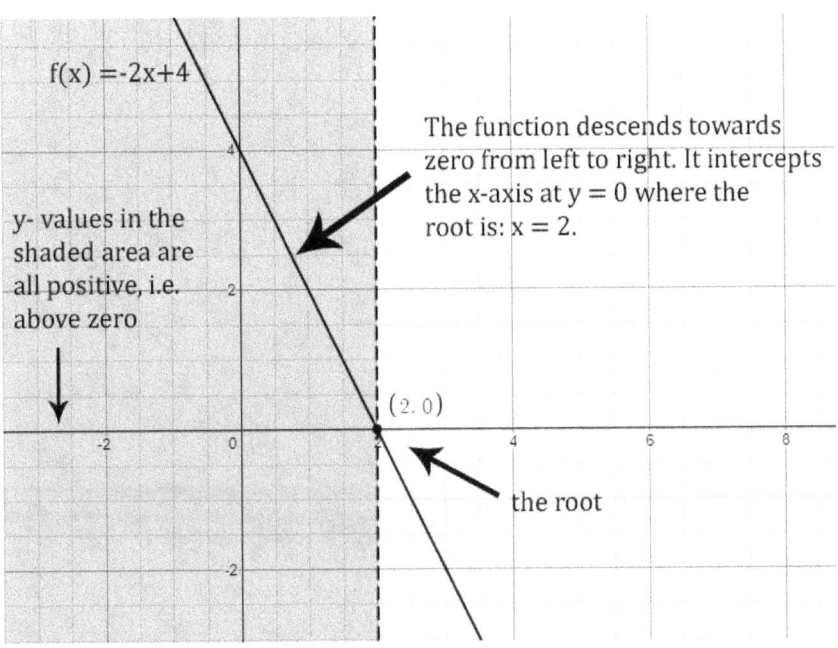

Figure 20.4, Desmos Link: https://sharpseries.ca/0/V1_L20.4.html

The function $f(x) = -2x + 4$ is a line in slope-intercept form, which is easy to plot as shown above. The x-intercept is at $(2, 0)$. All x values on the left of the point $(2, 0)$, i.e. in the interval $x < 2$, have y values that are positive. On the right side of $(2, 0)$ all y values are negative as the line dives under the x-axis at that position.

The x values in the grey area are all $x < 2$. This is a solution that is identical to the one found earlier algebraically.

Lex Sharp

Algebraic Rule 1

$$\frac{\frac{a}{b}}{\frac{c}{d}} = \frac{a}{b} \div \frac{c}{d} = \frac{a}{b} * \frac{d}{c} = \frac{ad}{bc}$$

Algebraic Rule 2

Fractions can be **simplified** as follows if the terms are connected by multiplications.

$$\frac{a}{b} * \frac{c}{a} = \frac{\cancel{a}}{b} * \frac{c}{\cancel{a}} = \frac{c}{b}$$

Alternately,

$$a * \frac{1}{a} = \frac{\cancel{a}}{1} * \frac{1}{\cancel{a}} = \frac{1}{1} = 1$$

or,

$$\frac{a}{a} = \frac{\cancel{a}}{\cancel{a}} = \frac{1}{1} = 1$$

Algebraic Rule 3

Extracting two fractions from an addition/subtraction of fractions:

$$\frac{a+b}{c} = \frac{a}{c} + \frac{b}{c}$$

$$\frac{a-b}{c} = \frac{a}{c} - \frac{b}{c}$$

This rule is often applied in both directions: left to right and right to left as well. In other words, learn to identify this relation as well:

$$\frac{a}{c} + \frac{b}{c} = \frac{a+b}{c}$$

$$\frac{a}{c} - \frac{b}{c} = \frac{a-b}{c}$$

Algebraic Rule 4

Any number can be expressed as a fraction by dividing it by 1.

$$Number = \frac{Number}{1}$$

Algebraic Rule 5

Any number a can be multiplied by 1, and 1 can be transformed into a fraction with any denominator b as follows:

$$a = a * 1 = a * \frac{b}{b} = \frac{ab}{b}$$

Algebraic Rule 6

Adding two fractions necessitates an identical denominator. These cannot be added until the identical denominators have been resolved.

Thus,

$$\frac{a}{b} + \frac{c}{d}$$

is not ready for the addition if $b \neq d$. First a common denominator must be found. Assuming that b and d are prime numbers, the common denominator is $b*d$.

Otherwise a **Least Common Denominator** (LCD) calculation must be employed first, which is not in the scope of this volume.

$$\frac{a}{b} * 1 + \frac{c}{d} * 1 =$$

$$\frac{a}{b} * \frac{d}{d} + \frac{c}{d} * \frac{b}{b} =$$

$$\frac{a * d}{bd} + \frac{c * b}{bd} = \frac{ac + cb}{bd}$$

Algebraic Rule 7

Multiplying fractions does not necessitate an identical denominator:

$$\frac{a}{b} * \frac{c}{d} = \frac{a * c}{b * d}$$

Additionally, in combination with *Algebraic Rule 4* the following can also be resolved:

$$a * \frac{b}{c} = \frac{a}{1} * \frac{b}{c} = \frac{a * b}{1 * c} = \frac{a * b}{c}$$

Algebraic Rule 8

$a^2 - b^2 = (a + b)(a - b)$

This formula is also known as the "Difference of Squares".

This rule is clearly stated with a difference of terms that are both perfect squares: a^2 and b^2. Nevertheless, non-perfect squares can also be used by applying the square root using the same pattern:

$m - n = (\sqrt{m} + \sqrt{n})(\sqrt{m} - \sqrt{n})$.

Algebraic Rule 9

Another variation on the "Difference of Squares" relation is:

$a^4 - b^4 = (a^2 + b^2)(a^2 - b^2) = (a^2 + b^2)(a + b)(a - b)$

Algebraic Rule 10

The following formula is known as the "Sum and Difference of Cubes":

$a^3 + b^3 = (a + b)(a^2 - ab + b^2)$, and

$a^3 - b^3 = (a - b)(a^2 + ab + b^2)$.

These formulas are sometimes collapsed into one shorter form as follows:

$a^3 \pm b^3 = (a \pm b)(a^2 \mp ab + b^2)$, and

Algebraic Rule 11

$(a + b)^2 = a^2 + 2ab + b^2$, and

$(a - b)^2 = a^2 - 2ab + b^2$.

These formulas represent perfect squares and are sometimes collapsed into one shorter form:

$(a \pm b)^2 = a^2 \pm 2ab + b^2$.

Algebraic Rule 12

Perfect cubes are represented by

$(a + b)^3 = a^3 + 3a^2b + 3ab^2 + b^3$, and

$(a - b)^3 = a^3 - 3a^2b + 3ab^2 - b^3$.

These formulas can be collapsed into one shorter form as follows:

$(a \pm b)^3 = a^3 \pm 3a^2b + 3ab^2 \pm b^3$.

Other Books in This Series

Lex Sharp

Lessons, Math 30 Diploma Prep

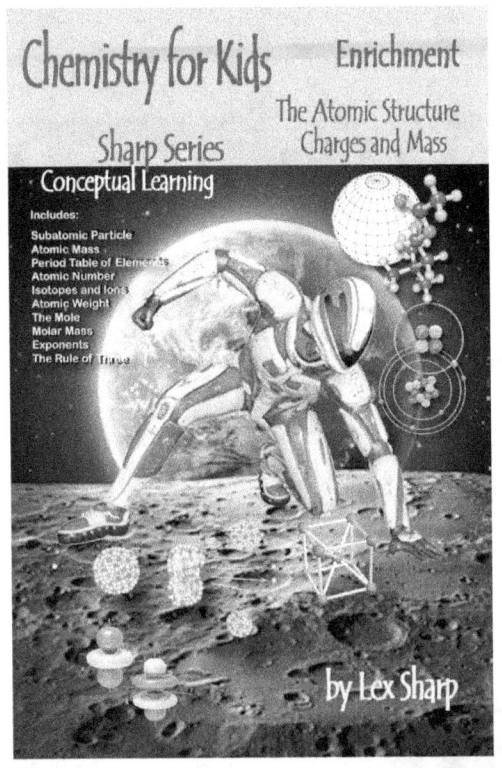

Lex Sharp

Lessons, Math 30 Diploma Prep

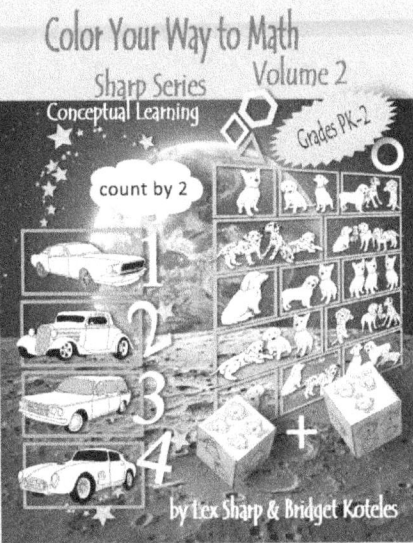

www.ingramcontent.com/pod-product-compliance
Lightning Source LLC
Chambersburg PA
CBHW050246220526
45465CB00002B/570